Global "Body Shopping"

An Indian Labor System in the Information
Technology Industry

Xiang Biao

PRINCETON UNIVERSITY PRESS
PRINCETON AND OXFORD

Copyright © 2007 by Princeton University Press
Published by Princeton University Press,
41 William Street, Princeton, New Jersey 08540

In the United Kingdom: Princeton University Press, 3 Market Place,
Woodstock, Oxfordshire OX20 1SY

Library of Congress Cataloging-in-Publication Data
Xiang, Biao.
   Global "body shopping": An Indian labor system in the information technology
industry / Xiang Biao.
      p.   cm. — (In-formation series)
   Includes bibliographical references and index.
   ISBN-13: 978-0-691-11851-2 (hardcover : alk. paper)
   ISBN-10: 0-691-11851-5 (hardcover : alk. paper)
   ISBN-13: 978-0-691-11852-9 (pbk. : alk. paper)
   ISBN-10: 0-691-11852-3 (pbk. : alk. paper)
      1. Electronic data processing personnel—India. 2. Labor mobility—India.
   3. India—Emigration and immigration. I. Title. II. Series.
   HD8039.D372I48   2007
   331.12'7910954—dc22   2006010309
British Library Cataloging-in-Publication Data is available

This book has been composed in Sabon with Futura display

Printed on acid-free paper. ∞

pup.princeton.edu

Printed in the United States of America

10   9   8   7   6   5   4   3   2

Global "Body Shopping"

FORMATION *Series*

Series Editor
PAUL RABINOW

A list of titles in the series appears at the back of the book.

Contents

Illustrations, Tables, Boxes

ACS	Australian Computer Society
ANZ	Australia and New Zealand Banking Group Limited
AP	Andhra Pradesh
APESM	Association of Professional Engineers, Scientists and Managers (Australia)
B2B	business to business
B2C	business to customer
BCA	Bachelor of Computer Applications
BPO	business process outsourcing
CAPSTRANS	Center for Asia-Pacific Social Transformation Studies
CBD	central business district
C-DOT	Center for Development of Telematics
CEO	chief executive officer
CTE	Commissionerate of Technical Education (Andhra Pradesh)
DCITA	Department of Communications, Information Technology and the Arts (Australia)

DES   Directorate of Economics and Statistics (Andhra Pradesh)

DIMA   Department of Immigration and Multicultural Affairs (Australia). DIMA was restructured into Department for Immigration and Multicultural and Indigenous Affairs on November 26, 2001.

ERP   enterprise resources planning

GRE   graduate record examination

HSS   Hindu Swayamsevak Sangh (Hindu Volunteers' Corps)

HUDA   Hyderabad Urban Development Authority

IDP   International Development Program (Australia)

IIM   Indian Institute of Management

IIT   Indian Institute of Technology

INS   Immigration and Naturalization Services (United States). INS was transitioned into the Department of Homeland Security as the United States Citizenship and Immigration Services on March 1, 2003.

IPO   initial public offer

IT&T   information technology and telecommunication

IT   information technology

ITAA   Information Technology Association of America

ITPA   Information Technology Professionals Association (Australia)

MBA   Master of Business Administration

MCA   Master of Computer Applications

MCSD   Microsoft Certified System Engineers

MIC   Malaysian Indian Congress

MIT   Massachusetts Institute of Technology

MSC   Multimedia Super Corridor (Malaysia)

MT   medical transcriptionists

Nasscom   National Association of Software and Service Companies (India)

NIIT	National Institute of Information Technology (India)
NRI	nonresident Indian
NTTF	National Technical Training Foundation (India)
OECD	Organisation for Economic Cooperation and Development
PCS	Patni Computer Systems
PIN	personal identification number
PQBS	pre-qualified business sponsors
PR	permanent resident
PRC	People's Republic of China
SBS	standard business sponsors
SCHE	State Council of Higher Education (Andhra Pradesh)
TCS	Tata Consultancy Services (India)
TOEFL	Test of English as a Foreign Language
UNICEF	United Nations Children's Fund (originally United Nations International Children's Emergency Fund)
USAID	U.S. Agency for International Development
VC	venture capitalists
VHP	Vishwa Hindu Parishad (World Hindu Council)
WTO	World Trade Organization
Y2K	year 2000

Currencies

AUD	Australian dollar. 1 AUD was 0.66 USD (28.58 INR) in the beginning of 2000, dropped to 0.48 USD (22.30 INR) in April 2001, and recovered to 0.52 (24.00 INR) in June 2001 when I left Sydney.
GBP	British pound. 1 GBP was 1.62 USD in the beginning of 2000, and 1.45 USD in the end of 2001.
INR	Indian rupee. 1 INR was between 0.021 and 0.023 USD during my fieldwork.

MYR	Malaysia ringgit. 1 MYR equaled about 0.26 USD.
SGD	Singapore dollar. 1 SGD equaled between 0.57 and 0.60 USD.
USD	U.S. dollar.

About "Lakh" and "Crore"

To be consistent with most documents in India, this book uses lakh and crore in measuring Indian rupees, where 1 lakh = 100,000, and 1 crore = 100 lakh = 10 million.

Prologue: A Stranger's Adventure

This book is a result of a research adventure by a Chinese-educated graduate student whose investigations focused on information-technology (IT) professionals migrating from India to Sydney, Australia, via a labor arrangement known (infamously) as body shopping. When I left China for the first time, in 1998, to pursue my doctorate at Oxford, I had no knowledge or even curiosity about IT, India, or migration to Australia. All I knew was that I *had* to work on a specifically non-Chinese case study. I made this decision because I did not want to be slotted into what clearly seems to be a division of labor in knowledge production in the international social sciences. While scholars from the West roam the world, researching and doing battle with "issues" in far-flung, non-Western countries, researchers from developing countries more typically specialize in "home" topics—as "local" scholars. My decision was supported unswervingly by my thesis supervisor, Dr. Frank Pieke. As a China specialist, Dr. Pieke holds that it is not enough for Western scholars to learn to understand China from a Chinese point of view and to encourage the "natives" to speak about themselves; instead Chinese scholars must develop their own *worldview* and thus become part of the international academic community as active critics.

But what "non-Chinese" topic should I choose? India was immediately appealing for two reasons: its cultural distance from China seemed sufficiently vast, and fieldwork on the subcontinent could excuse me from the burdens of learning another foreign language in addition to English. My concern with language also narrowed the choice to skilled migrants as the subject group. I chose Sydney as the case for migration destination because

I was told that the costs of living were lower there than in most major cities with sizable skilled Indian immigrants. It also seemed to me—instinctively—that Australia, as a relatively marginal Western country, *should* be more interesting than the much better known and studied United States.

Although I chose the subject based on practical calculations, I was obliged to come up with an elaborate theoretical rationale for my choice—not so much for my application for D.Phil. candidacy, but also because of my own anxiety about theories as a Chinese student. In the 1990s, when academic research as commonly understood became the mainstream activity of Chinese intellectuals after the waning of the intensely ideological periods of the Cultural Revolution and the post-Tiananmen Square incident, it was acutely felt that China lacked systematic social theories comparable to those in the West. The tendency to look to the West for theory was further encouraged by the widely shared desire for "joining the international (academic) track" (*yü guoji jiegui*). Researchers returning with Western degrees were almost worshiped. Numerous Western theoretical works were translated—often in a lamentably unreadable manner. ("Postmodernism" was translated—comically—as "mail-modernism" in a Chinese edition of Habermas.) For a young student like me, the obfuscation and obscurity of theoretical writings—even when properly translated—made it all the more mysterious and appealing. Thus, my Chinese friends and I envisaged my graduate-school training as a sort of a mission—like the legendary Monkey King's journey to the West—to seek the authentic sutra of anthropology.

Oxford, however, threw me into even deeper confusion and anxiety. The theoretical possibilities were unlimited, and jargon was everywhere. With my basic English I could understand only about one third of the sentences in seminars and lectures, which had the effect of making the repeated phrases and brilliantly constructed sentences sound all the more powerful—and overwhelming. After spending the entire first year reading (or, more accurately, scanning) and copying phrases frantically, I had cobbled together a report for my upgrading to D.Phil.-candidate status. My report proposed a project to delineate a "diasporic space" by comparing the "diasporic experiences and feelings" both at home and in the workplace of two groups of Indian migrants in Sydney: physicians and IT professionals. Although one of my examiners found the proposal "outrageous" and expressed this opinion twice during the defense I was allowed to go ahead with my fieldwork after the examiners concluded in their report that "little will be achieved to ask the student to rewrite."

I started my fieldwork immediately after arriving in Sydney in January 2000. Thanks to the Internet, I was able to contact Indian migrant associations in Australia beforehand, and began the process of recording different patterns of transnational networks and means of communications (i.e., their "diaspora experiences"). Although talking to people was much easier for me than reading books, the information that I was collecting was disparate and I was still anxious about theories and frameworks. One afternoon in February 2001 as I walked on the beach, too tired to think of how to patch the material together using the theoretical language of "multifaceted, multilayered" diasporic space, I let my common sense take charge and began turning over my informants' stories in my mind. It was as if I had emerged into a clearing. I rushed back to read through all the fieldnotes and by supper time had decided to shift my focus to IT professionals' mobility, that is, to the practice of body shopping. It is the linchpin connecting IT corporations, placement agents, small IT consultancies (body shops), and IT workers, and thus provided a solid basis for an ethnography. And the fact that body shopping had come into being for various *institutional* reasons—such as labor regulations and immigration control—struck me as highly significant. I did not know what the theoretical import of the study would be, but I simply could not focus on the "diasporic space" when a potentially rich case study loomed.

Body shopping was, however, a much more sensitive topic for detailed investigation. The relations between the workers and the body-shop operators, among Indian body-shop operators, and among workers, were fraught with sensitivity. "Did you visit other consultancies [as body shops referred to themselves] and what did they say?" "What did [the body shop operator who sponsored the worker] tell you?" were questions I had to face continually when the conversation turned to the subject. I usually described my interest in body shopping to the operators in terms of a research project on diasporic networks, and focused my questions on their business strategies. Thus I often appeared to the body-shop operators as an admirer of the glorious global success of Indian IT. By contrast, among the workers, with whom I often mingled without the body-shop sponsors' knowledge, I was more honest about my views. For example, I often encouraged the workers to sue the sponsors for breaching contracts. (When they replied by explaining why they could not take these actions, this helped me better understand their position.) Workers were also predisposed to trust me in light of my status as a temporary visitor to Australia of Chinese origin with no ties either to their employers or to any Western authorities. Indeed, in the view of many of my young Indian IT interlocutors, I was in the same marginal category as them ("*we* Indians and Chinese . . ."). Coming from China, I had no problem sleeping on the floor with five guys and innumerable cockroaches, and empathized with their anxiety about needing money and wanting to be generous at the same time.

Once focused on body shopping, I immediately noticed the high presence of the Telugu people from the state of Andhra Pradesh in southern

India among the IT professionals. (In fact, according to the state's finance and planning department (1999), Andhra Pradesh was home to 23 percent of all Indian IT professionals worldwide by the end of the 1990s.) But in the to and fro of the day-to-day business of body shopping, regional boundaries hardly mattered for my informants; what stood out were networks based on the common *nation* of origin. Thus, Telugu IT professionals are significant for this study not because of any special "cultural" features, but because they constituted the most representative sample to reflect the larger group of Indian IT people. For this reason, this book does not always mark the boundary between Telugu and other Indian IT people. By June 2001, at the end of the Australian leg of my field research, I had lived with thirteen Indian IT workers in four different places in the Sydney area. A few Telugu and Tamil workers became friends as well as key informants—in particular, Uday, Ashok, Rajan, and Venush (to protect the privacy of all respondents, I am withholding their real names). I had also engaged in open-ended interviews with 124 Indian professionals (mostly male and single, or married but alone in Sydney) and twenty-five Australian institutes (companies, professional organizations, universities, and government departments). Single interviews took one to two hours, mainly conducted at the respondent's place of residence; usually each respondent was interviewed more than once.

With a handful of contacts provided by my friends in Sydney, and pages of addresses of various institutes copied from Web sites, I set off for Hyderabad, the capital city of Andhra Pradesh, via Kuala Lumpur, where I stopped over to meet a group of Telugu IT professionals. In Hyderabad I conducted interviews with forty-three IT people, body-shop operators, government officials, and academics. I also spent ten days traveling between two towns and three villages in the West Godavari district in coastal Andhra, the region from which large numbers of IT professionals originated. My adventure came to an unexpected halt on September 11, 2001, when I collapsed with severe hepatitis. While the world reeled with images of the disaster in New York, I managed to board a flight to New Delhi, where I was looked after by my friend Tasnim Partapuri and her sister Zehra Husain's family—the public hospitals in Hyderabad being too crowded, and the private ones far too expensive. In Delhi, an otherwise agonizing time became a pleasant break that enabled me to collect considerable secondhand information, mainly from Indian media reports.

Despite the illness, my time in India was the most exciting and productive part of the adventure. It seemed as though everything I was interested in and concerned about was in Hyderabad waiting for me. Information technology and body shopping were the main things in the town. Unlike in Sydney, where both body-shop operators and workers were under constant stress and I had to navigate carefully, Hyderabad had a more relaxed environment. The large number of IT professionals on seemingly permanent stand-by to go abroad were happy to have a curious Chinese man accompany them on their jaunts around the city, having juice or cocoa and dropping into various body shops asking about the job-market trends in the West. Desire and despair were present, of course, but people tried to laugh at one another and themselves. It was in Hyderabad that globalization as a set of lived experiences full of ironies and contradictions—and as a powerful process of institutional change rearranging and reinforcing unequal human relations rather than merely intensifying connectedness—came fully into view. I was then able to conceptualize how to conceive of body shopping as a subcurrent of globalization that is constructed and experienced transnationally. My experiences in India also boosted my confidence in my chosen approach, which involved folding ethnographic observation into political economic analysis. This approach allows us to follow the flows and links of globalization—not only to document the literal cross-border connections and imaginations as much of the anthropological literature has attempted to do, but more importantly to bring into light what people are *not* aware of, such as the linkage between the pressure for increasing dowries in India and the global high-tech economy.

As I was writing my thesis, my parents in my hometown in southeast China, Wenzhou, told me that they had received a marriage proposal for me that came with an implicit promise of a substantial monetary "donation" to my family. Prior to my global adventures, I would have dismissed the proposal without much thought. This time, however, I wanted to know more. Although I was too shy to ask how high the bid was, I discovered that the proposal came from a well-to-do family, originally factory workers, recently made prosperous in the leather-shoe trade, and they wanted to reinforce their newfound status by finding a son-in-law with "international connections." Further queries revealed that the marriage age for at least some women in my region of China has decreased with the onset of rapid economic development, going against international experience and most projections. My experiences in India, including discussions with Indians about the institution of dowry, led me to initiate some lively discussions with friends in Wenzhou on recent changes to marriage and what they mean to Chinese society—a topic that otherwise might have been dismissed as trivial at a time when everyone was preoccupied by debates on WTO, corruption, and the rising cost of housing.

I might have written a different book had I studied a Chinese case comparable to body shopping. Such a book might have focused more on the operators' business strategies and how the practice can stimulate the larger economy in the homeland. This is because, quite frankly, I feel that I am more inclined to see human agency and potential in the Chinese case, whereas the facts of domination and inequality come more readily to mind

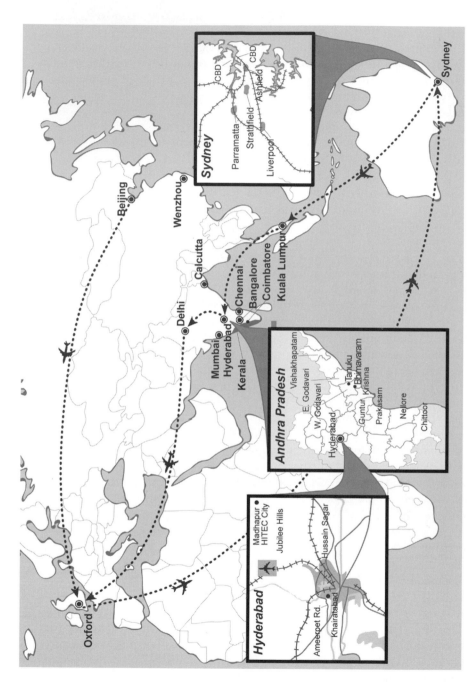

FIGURE 1. My Journey. Produced by Sarah Moser.

when I turn to India. Surely this difference has something to do with my *national* identity, which is of course political rather than cultural. I am neither proud nor ashamed of this, but am glad to be aware of it—thanks to my research for this book—and it certainly warrants more reflections on why I have adopted the different approaches. I believe that I have enriched myself more through this project than by studying a Chinese case debating either Chinese or Western literature. I also believe that Chinese society will be better understood if more Indian and other scholars could work on China. I thus hope this book will contribute in some modest way to an increase in South-South dialogue in anthropology.

Acknowledgments

Without the generous support from many individuals and organizations I would have not been able to complete this project and the book. Apart from my D.Phil. supervisor Dr. Frank Pieke, my co-supervisor, Dr. Steven Vertovec at Oxford, gave me valuable instruction on the research. Laurence J. C. Ma, Professor Emeritus of Geography of the University of Akron, in Ohio, has advised me on both research and career development. During my fieldwork in Australia, I was affiliated with the Centre for Asia-Pacific Social Transformation Studies (CAPSTRANS), University of Wollongong. Dr. Robyn Iredale and her family, Professor Tim Turpin, and Mr. Martin Smith gave me both intellectual advice and emotional support. Professor Stephen Castles, then director of CAPSTRANS, shared with me his insights on international migration. Other CAPSTRANS members, among them Ms. Maureen Dibden, Dr. Mark Rix, Mr. Layton Montgomery, Dr. Fei Guo, Dr. Adrian Vickers, Dr. Jan Elliot, Dr. Tana Li, Professor Ken Young, and Mr. Matt Ngui, all made me feel at home. Professor Bhaskar Rao of the University of New South Wales and Dr. Joseph Davis of the University of Wollongong helped me establish contacts with Indian IT professionals in Sydney. Apart from my Indian friends, I also occasionally stayed with the Monro family and Li Nan's family in Sydney. I deeply enjoyed the warm home atmosphere. Dr. Diana Wong of Universiti Kebangsaan Malaysia accommodated me in Kuala Lumpur on my way from Sydney to Hyderabad, and I learned a great deal from conversations with her.

In Hyderabad, Col. M. Vijay Kumar of the Software Technology Parks of India, Hyderabad; Mr. Randeep Sudan, Special Secretary to the Chief

Minister of Andhra Pradesh; Mr. P. T. Prabhakar of the Andhra Pradesh State Commissionerate of Technical Education; and Dr. T. H. Chowdary, Information Technology Advisor to the Government of Andhra Pradesh, to name a few, helped me with the research. I am particularly grateful to Dr. R. Siva Prasad at the Department of Anthropology, University of Hyderabad, for sharing with me his sharp insights into Andhra society. I also benefited greatly from conversations with staff and students at Osmania University and the Centre for Economic and Social Studies in Hyderabad.

I thank my family for their unconditional support of my work. I thank K. C. Wong International Education Foundation and St. Hugh's Graduate Bursaries, which made my study at Oxford possible. My writing of the thesis was financially supported by the Frere Exhibition for Indian Studies Funds. I thank my friends Leila Fernandez-Stembridge, Antoine Pecoud, and Tasnim Partapuri for their constructive comments on the thesis. My examiners Professors Peter van der Veer at University of Utrecht, the Netherlands, and Stephen Castles who had moved to head the Refugee Studies Centre at Oxford by that time, offered me valuable suggestions for revision.

Developing the thesis into the book was far more difficult than I had anticipated. I express my special gratitude to my friend Vani S. in Singapore, who revised the manuscript extensively. Her extreme generosity turned the revising into a rewarding learning experience for me. I did part of the rewriting as a postdoctorate fellow at the Asia Research Institute, National University of Singapore, and I am grateful for the institute's financial support for the revision. I thank Fred Appel at Princeton University Press for his patience and help. My friend Sarah Moser produced figure 1. Routledge publishers and the journal *International Migration* kindly permitted me to include in the book material that I had earlier published with them.

Finally, my sincere gratitude goes to my informants, who shared with me their hopes, happiness, and struggles. Some of these became dear friends, among them Vandrangi Siviram, Rajesh Venu-Gopal, Manoj Joshi, Srinivas Dseenu, Sudhaka Reddy, Vandrangi Kishore, Sayeed Unisa, Aparesh Patra, Laxman Konduriven, Rakesh Jaggi, Gajendra Pareek, and Rishy Pareek. One of my tactics for collecting information was to tell them, when they asked me to hang out with them in the evening, that I would continue working unless they answered some tricky questions such as about the relationship between a certain body shop and a friend of theirs. They also constantly asked me difficult questions, such as when I would be married ("In short order!"), and more commonly, "Are you coming back to Sydney/Hyderabad soon?!" Yes!

Global "Body Shopping"

This ethnography is about embeddedness and disembeddedness—about how new human connections and disconnections are created and ultimately contribute to a process of abstraction in global capitalism today.

Consider the following. Flying in the face of an industry projection of a labor shortage of 850,000 in the information-technology sector in the United States for the year 2001,[1] the first eight months of that year saw more than 350,000 high-tech workers, mostly in IT, laid off,[2] a figure climbing to 600,000 by November.[3] And in another twist, the period 1998–2000 found delegations from more than twenty countries coming to India to recruit IT workers,[4] but in May 2001, some 50,000 Indian "computer whiz-kids" in the United States were reportedly jobless.[5] These scenarios immediately prompt at least two questions: How is skilled labor managed internationally to serve an extremely volatile global IT market? And what are the experiences of these skilled migrants in the so-called New Economy,[6] where "risk, uncertainty, and constant change are the rule, rather than the exception" (Atkinson and Court 1998, 8)? These concerns impinging on the livelihood of overwhelming numbers of migrant workers also underscore the relevance of a more general area for critical inquiry, namely: How are social relationships being restructured in response to economic globalization?

This book presents a configuration of the India-based, global labor-management system in the IT industry known as "body shopping," focusing specifically on its operations through Hyderabad, India, and in Sydney. Spanning the period 2000–2001 that immediately followed the frenzied demand for "Y2K" software fixes worldwide (to prevent computer systems from mistaking the year 2000—coded as "00"—for 1900), my narrative traces how this volatile global industry is constructed through concrete human relationships. My study gives centrality to labor—a much neglected or even deliberately omitted dimension in public discourse about the New Economy and its crucial underpinning IT—and shows how global high-tech hubs, such as the iconic "Silicon Valley" of Palo Alto in northern California, are intimately connected to women and children in rural India through the processes of IT labor production and surplus appropriation. Thus, this account constitutes a "global ethnography" not just in the sense of documenting how people behave transnationally, but also in clarifying how different regions of the world are related to each other institutionally and structurally.

Related to this perspective, the current round of economic globalization, which started in the 1970s, is understood as a continuation of a process of abstraction that has been central to the evolution of capitalism. Undeniably, the process of abstraction whereby markets *dis*embed from familial, religious, and communal relationships to become an autonomous and dominant social force has gathered considerable momentum since the 1970s. The flexibilization of the labor market, economic deregulation and decentralization, and the globalization of financial markets with the revolution in IT, in particular, have significantly freed the market from tangible social relations and from the primary institution managing public life, the nation-state.[7] The bewildering abstraction in the global New Economy is well portrayed by Castells (1996, 474) as:

> ultimately dependent upon the nonhuman capitalistic logic of an electronically operated, random processing of information. It is indeed capitalism in its pure expression of the endless search for money by money through the production of commodities by commodities. But money has become almost entirely independent from production, including production of services, by escaping into the networks of higher-order electronic interactions barely understood by its managers. While capitalism still rules, capitalists . . . prosper as appendixes to a mighty whirlwind which manifests its will by spread points and futures options ratings in the global flashes of computer screens.

Besides the abstraction of economic practice, there is the conceptual abstraction resulting from the ascendance of formal economics (and its professionalisms) both in the social sciences and in public discourse (Polanyi 1957b; see also Carrier 1998a). Neoclassical economic thinking has become so pervasive in the real world that "virtualism," as Carrier and his associates (1998) argued, has become a de facto feature of economic life. Policy-making and economic regulation increasingly conform to abstract, mathematically derived, and virtual (presumed universal) economic "laws"

2

rather than responding to observations of the ground realities—actual people's real needs. Institutions such as the World Bank, the International Monetary Fund, and the mushrooming master of business administration (MBA) courses worldwide have been the backbone of this process of (professionalized) abstraction. Neoliberal[8] thinkers no longer need worry that their prognostications are criticized as ignoring stark ground realities— they are inventing a new one.

Yet, despite the powerful trends of abstraction and virtualism in economic practice and thinking, anthropologists have primarily been trying to make sense of the world by emphasizing "embeddedness," focusing on how economic activities, no matter how abstract and global, *still* depend on and are still shaped by concrete human connections (e.g., Eriksen 2003). For example, in studies of transnational migration—an important dimension of globalization—much of the existing anthropological and sociological literature has explained it centrally by the existence of "networks" in which migratory flows are said to embed (Boyd 1989; Brettell 2000; Moretti 1999; Portes 1995; Tilly 1990; Zahniser 1999; for a recent critical review, see Krissman 2005). Indeed, according to Douglas Massey (1990; 1994; Massey et al., 1987; Massey et al., 1993; Massey et al., 1994), one of the defining figures in migration research of the last two decades, migration is so strongly embedded in and determined by networks that it would become progressively independent of the original macro-socioeconomic causes and eventually come to a stage of autonomous existence. Certainly such insights provide valuable correctives to the neoclassical universalistic view of society and atomized view of actor, but there is a danger here of losing sight of the overall trend of social change. For most people, the real pressing questions concern why and how their society is changing so fast, rather than what has *not* changed. People may need to be told that ethnic networks still matter in migration, but they are keener to know, say, why IT professionals were constantly on the move and why they made a fortune by creating nothing but Web sites. By emphasizing "embeddedness," anthropologists and sociologists have perhaps asked the wrong question in the first place about how economic activities—as though imposed from outside—are inserted into social relations. A more fruitful question may be exactly the opposite: how people *develop* social relations—seen as a holistic process of which their economic activities are a part—that lead to economic globalization.[9] A point here is: Are anthropologists up to the task of explaining such phenomena as the emergence of the stock market?

This book attempts to take up this challenge. On the one hand it demonstrates that the process of abstraction is by no means an inevitable consequence of "economic laws," but is constructed and sustained through the rearrangement of various institutions, the interplay between unequal

3

socioeconomic relations at different levels, and the establishment of particular ideologies. On the other hand the book stresses that while abstraction is strongly conditioned by established institutions (in the present case ranging from caste and dowry in India to labor and immigration control in the destination country), it should not be seen merely as a process of embedding. For instance body-shopping operations were based on networks, but the networks, far from appearing to embed Indian IT professionals' mobility, render it with more uncertainty, precisely by facilitating multiple, global, and multidirectional movement to escape as well as exploit economic volatility, and have thus overall contributed to a process of *dis*embedding rather than embedding. Before turning to the dynamics internal to body shopping (summarized as ethnicization, individualization, and transnationalization), it is necessary to establish the basic characteristics of this labor-management system against a historical and institutional background.

Body Shopping: Brief Overview

Body shopping is arguably a uniquely Indian practice whereby an Indian-run consultancy (body shop) anywhere in the world recruits IT workers, in most cases from India, to be placed out as project-based labor with different clients. Unlike conventional recruitment agents who *introduce* employees to employers, body shops *manage* workers on behalf of employers—from sponsoring their temporary work visas to paying their salaries, arranging for accommodation and the like. Thus, workers do not enter into any direct relationships with their contract employers and can be retrenched at any time, whereupon the body-shop sponsor either is able to place them out to a different client or puts them on the bench to await a placement. Acting in association, body-shop operators link up with each other in the same region or in different countries, sending IT workers to where they are required. Although it is almost impossible to accurately estimate the extent of this global business, it is enormous. At any given time during 2000–2001 there were perhaps over one thousand agents specializing in the supply of temporary Indian IT workers across the United States and hundreds in northern California alone,[10] and these agents were managing as many as 20,000 IT workers in the United States.[11] During my fieldwork, most of the Indian informants estimated that no less than thirty-five body shops were managing more than 1,000 Indian IT workers in Sydney in late 2000.

Quite the opposite of what is usually assumed, software development and services are highly labor intensive, particularly at the phase of programming (coding a software design into computer languages) and testing, or debugging (removing errors in program design). Most of the Indian IT workers migrating through body shopping took on such tedious, unrelentingly monotonous, and low-paying "donkey work." Hence, some informants suggested, the term—made up of "body" (rather than "brain"), indicating the labor-intensive nature of the work, and "shopping," which implies quick and easy purchases—in contrast with what is conveyed by the phrase "head hunting" used for senior IT positions and in other professional recruitment. To reflect this aspect of their work and more importantly their position in the labor market, this study refers to those managed by body shops as "IT workers" although they and the body-shop operators called them "consultants" or "IT professionals."

It is important to clarify at the outset that the global labor supply and management scheme of my study is not synonymous with an earlier, officially endorsed and similarly tagged practice in which India-based companies sent their staff to provide on-site software services for overseas clients, and these employees, who received an overseas allowance on top of their regular salaries, returned to their offices once the project was completed. The term "body shopping" first surfaced in this context with the establishment in 1974 of Tata Consultancy Services (TCS) in Mumbai, India's first export-oriented software-service company.[12] The body-shopping practice of my study developed during the late 1990s and is distinguished by quite different terms and practices. First, the ubiquitous practice of "benching" workers: quite simply, IT workers sponsored to enter a destination country on a temporary work visa without any prior job opening were "put on the bench" upon their arrival and subsequently, between job placements, without being paid or given a nominal stipend. Second, body shops—the term I use exclusively for the Indian consultancies engaged in the IT labor recruitment since the late 1990s—functioned in association with a chain of placement agents. This was largely because big corporate clients now outsourced their labor-management tasks to a single or a very limited number of large placement agents only, and body-shop operators, invariably small players, thus had to secure job openings through other placement agents who liaised with, often through yet another layer of agents, large corporations. Each agent in the chain took away part of the worker's monthly wage as part of the deal.

Whereas the earlier on-site services emerged primarily as a response to the labor shortage in the West, benching and agent chains are outcomes of various newer developments in the global IT industry and beyond. The widening application of Internet technologies enabled corporations to deterritorialize their production and management to an unprecedented extent, producing huge demands for IT workers. More importantly, since software packages had to be customized for the specific needs of different projects, it became necessary for IT professionals to move from one on-site

5

project to another. The "financialization" of the high-tech industry in the late 1990s—whereby industry ups and downs were determined by stock-market impulses—made large-scale firings and hirings an everyday event. The IT industry has thus needed not only a sufficient supply of skills but a mobile workforce so that it can respond to market fluctuations with minimum time lag. The need for the Y2K fix toward the end of the 1990s spurred a tremendous expansion in body shopping and further reinforced the practice of benching, which became entrenched during the following dot-com boom.

The specialized recruitment of Indian IT workers through body shopping also operated against a deep-rooted institutional background. On the one hand governments in developed countries have been rationalizing their immigration policies to facilitate the immigration of IT professionals since the early 1990s, such as the U.S. H-1B visa introduced in 1992, which allows foreign professionals to work for three years, renewable up to six. (Well over half of the H-1B visa holders were IT professionals.) But on the other hand governments still impose certain restrictions on companies in order to protect local labor and minimize any possible burdens on the state welfare system. In Australia, for example, it is technically illegal to sponsor the entry of foreign workers without confirmed job openings and for the sponsors to not pay these workers even when they are not working. Meanwhile, however, IT corporations require a smooth flow of immediately available short-term skilled labor. Body shopping removed the friction caused by state regulation by circumventing it: the benching practice of bringing in workers beforehand enabled corporations to select and dispose of workers anytime; agent chains freed large IT companies and the placement agents handling their labor management from obligations under the labor laws. The arrangement of agent chains made a highly externalized labor market comprising countless individual workers and brokers quite manageable from the industry's point of view. Furthermore, since the migrant workers shouldered the main part of the costs of relocation and of being benched, body shopping did not undermine the interest of the state or society in the receiving country.

The two forms of body shopping, despite the differences, appear almost identical in formal documents like visa applications. Except for a small proportion of IT professionals traveling on business visas for company projects overseas, most foreign workers must be sponsored by companies in destination countries to enter on a temporary work permit. As with client-company sponsors for the Indian consultants sent overseas, body-shop operators always claimed on paper that they were technology companies hiring workers for their own projects. This was not always a false claim. It is important to stress that a body shop, particularly in the destination country, was often a hybrid of a labor-placement agent, a software-services provider, a software-development house, and sometimes an IT training institute.[13] They also used benched workers for their own software development or training from time to time, and superficially this made the workers look like their "employees" who might then also be assigned to their "clients."

The hybrid nature of body-shop consultancies lent them great leeway in operating transnationally and differentiated body shopping from other international contract-labor migration schemes, in particular those for guest workers in western Europe in the 1960s and 1970s, and from project workers and contract laborers in Middle East countries and other regions. By being registered as high-tech firms, body shops were not held to the regulations imposed on registered labor agents, whether in India or the receiving countries; and by projecting the workers sponsored overseas as members of joint projects, they also circumvented some migration regulations. Thus, body shops not only brought workers from India to Australia, say, but also placed workers in a third or fourth country via Australia. Multiple mobility between various countries over a short span of time, rare for other migrants, was common for Indian IT workers. Body shopping matched mobile labor to volatile capital and was deeply integrated into the global New Economy.

Ethnicization, Individualization, and Transnationalization

At first glance, the body-shopping practice fits well with the anthropological embeddedness narrative of globalization: it is global yet ethnicized. The business remained an "Indian" phenomenon, not found among other sizeable groups of, for example, migrant Chinese, Filipino, and Brazilian IT professionals. In fact, the scale and reach of body-shopping operations had in a way led to an "ethnicization" of the entire global IT labor force: Indians constituted 74 percent of all computer-related H-1B visa holders in 1998–1999 in the United States (INS 2000) and 78 percent of all the foreign IT professionals entering the United Kingdom in 2002 (Clarke and Salt 2003, 572). The H-1B visa has even acquired the moniker "Indian visa." One informant told me that when one of his friends applied for a job as an accountant in the United States and asked why she was instead given a job as a computer programmer, the answer was: "You are Indian, you can do this."

The ethnicization of body shopping, however, had little to do with "ethnicity" as it is usually understood, since Indian Hindus, Muslims, Sikhs, and Christians were basically treated the same in body shops. Ethnic networks were important insofar as the cultural attributes they carried enhanced the assurances of complicity that enabled body-shop operators to get around state regulations—yet keep up the appearance of operating within them—and to control or expect compliance from the workers they sponsored. On occasion, body-shop operators in Sydney brought in non-Indian workers (mostly Filipinos) only if there were genuine job openings waiting for them; and when these workers were laid off they were not put on the bench without pay but received partial salaries, as if they were on vacation or medical leave. One body-shop operator in Sydney sent a few white Australian IT workers to the United States but did not repeat the attempt after they refused to share bedrooms with the Indian workers. When I asked both workers and body-shop operators why body shops managed Indians only, most replied that this was "a professional choice": Indians were simply the world's best in IT! Thus, even in rhetoric, it is perceived professional excellence, rather than cultural distinctiveness, that served as a base for the process of ethnicization.

Ethnicization is also related to the dynamic inherent in body shopping. Entrepreneur-aspiring IT workers accepted the conditions of migrating through body shopping not only because this was seen as the first step for entering the global market, but also because they often started accumulating initial capital by acting as subagents or by setting up body shops themselves. Subar, a young Telugu who went to Sydney through a body shop in 1999, had "something burning in the heart, a sort of anger" to run a business of his own and had registered two companies by early 2001, one of them for body shopping. Body shopping was regarded a particularly feasible starting point not only because it required low investment, but more importantly, because it facilitated other IT businesses by generating quick cash flows, developing client connections, and providing free (benched) labor. While earlier body shops ceased the business after upgrading to become fully-fledged technology firms, new ones were set up bringing more Indian workers to the global market and perpetuating body shopping as an "ethnic" business.

In examining the internal mechanisms of so-called ethnic economies, existing literature has pointed to the collective social forces in that particular ethnic community (for example value introjection, reciprocity transactions, bounded solidarity, and enforceable trust; see Portes and Sensenbrenner 1993).[14] However, accompanying the ethnicization of the body-shopping business are not the forces of a bounded collectivity, but rather a process of *individualization*—the tendency among Indian IT workers to size up their surrounding society and make decisions on individual rather than collective terms. Taking "workers" as one category and talking about "workers versus sponsors," as I did during interviews in the beginning of my fieldwork, was seen as naïve. "Everyone is different," was, I found, the view that practically everyone shared. It was a common practice that before being placed for the first time, workers were not paid but that between subsequent placements they might be given a stipend, even though in both cases they did not bring earnings to the body-shop operator. The reasoning was that until getting their first placement, workers had yet to prove their "merit," and, therefore, the body shop was not responsible for their unemployment, but once having proved their employability, workers were treated differently. Once again this was projected as part of "professionalism": people should be treated solely according to merit—as validated by the market, of course.

This individualization did not necessarily imply any change in the basic conceptions of personhood, nor did it make Indian workers more "individualistic" in the sense described by Western social theorists.[15] Most of my informants took particular pride in maintaining "Indian culture" (being family oriented and religious) while at the same time succeeding in the global market. Individualization in the case of my informants was less about "self" and essentially about the perception of society—particularly how one should apprehend uncertainty. Central to this perception is the belief that market flexibility and uncertainty made personal merit the key for individual success; group solidarity or group-based conflict of interests (e.g., employer versus employee) were not relevant. Indeed, benched workers would see benching as due to market "mismatch," a lack of "merit," or just plain luck, for which their sponsors were not to blame. The individualism of the Indian IT workers also justified hierarchic differentiation as natural, because every individual has different merits, which was again crucial for maintaining body shops' control of workers.

Owing to the process of individualization, the ethnicization of body-shopping operations did not generate the negative effects of collectivity such as the leveling pressures that discourage members from becoming too successful in the mainstream society (Portes and Sensenbrenner 1993). Far from that, body-shop operators encouraged workers to move up. If a worker left his sponsoring body shop to join another he would be threatened, but if a worker left after becoming a permanent resident (PR) in Australia (and no longer required a sponsor) or finding a job in the United States, even before completing the contract period with the sponsoring body shop, he would be congratulated. Becoming an Australian PR or going to the United States was considered a winning promotion, and, with successful examples in their fold, a body-shop operator could expect to attract more good new workers. Those who moved up, in turn, helped to expand the body-shopping networks worldwide as subagents. In short, individualization individualizes risks—that is, it disperses risk to individual workers to the benefit (profit) of body-shop operators—even as it individualizes opportunities for upward mobility, enabling the body-shopping scheme to continue expanding.

An important condition for the intertwining of ethnicization and individualization is the process of *transnationalization*. Transnationalization of body-shopping operations was impelled by IT workers' desire for personal career success, for which multiple transnational mobility was almost indispensable. In order to recruit highly qualified IT workers from India, a manager at Mastech, a large Indian-run global IT service company headquartered in the United States,[16] told me: "We have to emphasize that we can provide *global opportunities*. After working one year in Australia, we can send you to Canada, America, or Europe." But why are *Indian* IT professionals in particular globally popular? Behind the obvious reasons, such as high qualifications, low wages, and large numbers, is the high *transnational* surplus value that they embody, which can be measured as the disparity between the input for producing IT labor in India and the prevailing wages in the global market. The ethnicization of body shopping is thus fundamentally a reflection of an international division of labor where India specializes in producing (even overproducing) IT labor. The transnationalization of body-shopping operations enabled Indian IT workers to pursue individual success in the disembedding global market, but was itself situated within the established international economic order. Thus the intertwining of individualization and ethnicization.

For this reason, the transnationalization of body shopping by no means implies the emergence of autonomous "transnational spaces," as some literature of transnationalism suggested (e.g., Pries 2001).[17] What can be ascertained is instead a "world system of body shopping"—a system comprising boundaried nation-states that occupy differential positions in IT workers' migration strategies, and that assume mutually complementary functions in the global body-shopping business. This world system of body shopping was dominated by two centers: India and the United States. Before moving on to a new country from Australia, Indian workers went back to India almost as a matter of routine to attend courses to upgrade their skills and consult body shops there for information on hot opportunities in the world labor market. And it was often body shops and IT firms in India that put their counterparts in Australia in contact with those elsewhere. India was thus not only a source country of flexibilized IT labor, but also a coordinating center for global labor mobility. The United States stood as the major destination point and as the reference standard when IT workers compared the pros and cons of possible destinations that might lead them there. While India contributed the most valuable assets to the global IT industry, the United States attracted ready-made, highly qualified IT professionals from all over the world. Australia, along with Canada and the United Kingdom, served as bases of security that were essential to sustaining the hypermobility because markets there were less volatile, welfare policies generous, and applications for PR status met with relatively high success. Apart from these countries, there are gateway destinations that were seen as entrances or springboards to the global market, including Malaysia, Singapore, Hong Kong, and some Middle East countries; the satellites of the U.S. market, namely the Caribbean and Latin America; and finally, the new frontier zones of non-English-speaking industrialized markets that held out some promise for significant numbers during the slowdown, particularly Germany, Japan, and South Korea.

In this world system, Indian IT workers' seemingly expansive, multidirectional flows and networks have in fact deepened the concentration, rather than dispersing of wealth, power, and influence to certain loci and groups. The transnationalization of body shopping is important not because it created new cross-border flows and links, but because it embodied new strategies for wealth creation, new means of value transfer, and new forms of inequality. The process of abstraction, represented by the intensified transnational mobility of capital and labor in the present case, is to a great extent created by *and* sustains the global status quo.

Structure of the Book

In order to reveal the internal workings of body shopping and its broader institutional significance in an integrated manner, this book is structured in the following way. Chapter 1 provides the institutional background of body shopping. Chapter 2 documents how an astonishing amount of resources was channeled to the production and overproduction of the IT labor force in Andhra Pradesh, particularly through the institution of dowry. Chapter 3 demonstrates that based on the massive labor pool, body shops in India were not only able to provide especially cheap and flexible labor to the global market, but lived on the workers by charging them various fees and utilizing their free labor to survive the market slowdown.

From chapter 4 on, the book details the business operation of body shops. Chapter 4 traces the developmental trajectories of body-shopping businesses in Sydney, particularly how they evolved in overlapping with other IT operations. Chapter 5 focuses on the structure of agent chains in which recruitment agents of different sizes depended on each other by assuming different functions in dealing with the market, the state, and the workers. In these chains, body shops carved out their niche by benching. Then how did the body-shop operators elicit compliance from the benched workers? Chapter 6 suggests that the answer lies in various sets of relations: those between body-shops operators and the larger Indian

11

community in Sydney, between body shops, workers, and middlepersons, and among workers themselves. Equally, and possibly more important are workers' career strategies, which made it a vested interest for some workers to reinforce the body-shopping practice.

Chapter 7 asks the question: does body shopping indicate any fundamental change in our world? By mapping out the world system of body shopping, this chapter also clarifies the position of India-Australia body shopping in the global picture. The book ends with a discussion of the relation of body shopping to the much celebrated "IT miracles" and the local society in India. I suggest that India's IT success and body shopping are partly a result of intensified inequality at different levels: the macro (the international economic order and the socioeconomic structure in India), meso (relations between IT companies of different sizes), and micro (within a body shop). This is also the underlying logic of the process of abstraction.

Chapter 1
The Global Niche for Body Shopping

The single most important catalyst defining the form of the body-shopping practice was the global demand for the "Y2K" programs. Primarily based on the largely obsolete IBM mainframe technology, the Y2K programs involved little innovative design, but their implantations were extremely labor and time intensive. Hence, it made more sense for companies to outsource their Y2K projects to software service firms (vendors) who were better placed to organize large numbers of contract workers dedicated to this task. When the Y2K frenzy became widespread by 1998 in Australia, larger vendors often targeted clients with a high revenue base and willing to consider broader involvements beyond Y2K projects, some even setting their sights exclusively on Fortune 500 clients. This led to a flurry among small-sized vendors rushing in to meet the needs of the mass of smaller clients. Small vendors could not of course afford to set up branches overseas to recruit workers and instead collaborated with recruitment agents in labor-sending countries—mostly India in this instance. Soon afterward, these small vendors also provided workers to the large software vendors apart from to clients. For example, in 1997 ICON Recruitment Pty. Ltd., one of the largest IT placement agents in Australia, set up a liaison office in Bangalore, Karnataka, a state in southern India, and worked with over ten firms all over India to recruit Y2K programmers. Then, in 1999 when the demand for Y2K programmers started scaling down and the numbers of Indian labor vendors in Sydney had increased, ICON closed down its Bangalore office and recruited workers through Indian-run body shops in Australia that were bringing in IT workers through their associates in India, which was more cost effective and efficient for

ICON.[1] This multitiered and network-based market structure gave rise to agent chains, which became a defining characteristic of body-shopping operations in the United States by 1996 or 1997 and a commonplace in Australia after 1998. The body-shop practice of benching workers developed around 2000 in both Australia and the United States, after the Y2K heat was off and large numbers of IT programmers sat idle, awaiting new job opportunities.

This, however, does not mean that the body-shopping practice was an accidental by-product of the Y2K programs: the demand for a highly flexible international labor force in the IT industry had been built up well before; and the body shopping business was consolidated in the post-Y2K phase, specifically during the dot-com boom years. Thus we must examine the development of body shopping in a much larger context.

"Financial Democracy" and the Virtual Shortage of IT Labor

A high level of labor mobility—the key precondition for the emergence of the body-shopping business—was not always a necessary element of the IT industry, but one specific to the contemporary stage of development driven by Internet technologies, and in which the links with globalized capital played no small part. Internet technologies since the 1990s have greatly facilitated the management of resources on a global scale.[2] Internet-based "B2C" (business to customer) and "B2B" (business to business) transactions link together consumers, material suppliers, manufacturers, and dealers in real time, regardless of where they are, to reach an optimal arrangement of production and management. It is thus no surprise that the earliest major commercial investors in Internet-related IT services were from the finance sector—the most globalized fraction of capital—and they are still among the largest consumers. Indeed, many, including the Australian minister for communications and IT Richard Alston (1999a), exulted that IT had brought about "financial democracy": "In 1999, everybody can be in Wall Street."

Furthermore, Internet commerce, or "e-commerce" more broadly,[3] was arguably leading to a "tariff-free" economy and thus fulfilling the true spirit of capital to maximize itself by circulating freely.[4] Set against projections that global e-commerce would make up 1.3–3.3 percent of global GDP by 2001—equivalent to three times the size of Australia's economy—e-commerce in Australia was predicted to increase to AUD 1.3 billion in 2001, from AUD 61 million in 1997 (Alston 1999b). Buoyed by projections, the IT industry became the top investment priority in the 1990s worldwide, particularly through the Nasdaq Stock Exchange, which was the first fully "electronicalized" and thus "deterritorialized" exchange and had surpassed the New York Stock Exchange as the largest in the United States. The amount of capital injected into the IT sector was breathtaking: a USD 100 investment in Dell Computer made in 1988 multiplied in value to USD 56,470 by January 1999 (Sklair 2001, 263). The Street.com Internet sector index measuring IT stock performance increased by 164 percent over the year 1999 (compared to 19 percent growth of the Dow Jones Industrial Average and the Standard and Poor's 500). The years 1997–99 were thus called the "Net years" in the finance world.[5]

Besides stock-market investors, the IT sector attracted funds from venture capitalists[6] and angel investors. Venture capitalists invested in a startup's business plan for a share in profits, but expected their real killings from the start-up's initial public offer (IPO). In this sense, therefore, had there been no Nasdaq, there surely would not have been so many IT venture capitalists. By comparison, angel investors were usually IT industry veterans who made much smaller investments and who supervised the business closely, often contributing their technological expertise and business connections. In many cases, angel investors and dot-com start-ups would together solicit capital from venture capitalists when the prospects of their business plan became clearer. Considerable numbers of established Indian IT professionals in Silicon Valley often dubbed "gurus" mentored young Indian IT entrepreneurs in the United States as angel investors. Venture capitalists and angel investors pumped in about USD 185 billion into the high-tech sector worldwide in the year 2000 (Nasscom 2000).

The large injections of capital from various sources created tens of thousands of new IT jobs and brought about labor mobility. But this tells only part of the story. Beginning from the late 1990s, industrial bodies in all the major developed countries produced alarming estimates of IT labor shortages. For example, an Australian government taskforce reported a shortage of 31,500 in 1999, and projected shortfalls of 89,300 for the period 1999–2002 and of 180,200 by 2004 (IT&T Skills Taskforce 1999). In the United States, even with the ongoing freeze in recruitment and IT labor redundancies worldwide in mid-2001, the Information Technology Association of America (ITAA) still projected 425,000 unfilled IT job vacancies for 2001–2002 (ITAA 2001). Furthermore, major software companies in the United States had formed powerful lobbying groups to push for liberalizing migration schemes for foreign IT labor, making political-party campaign contributions for this purpose.[7]

The widely broadcast warnings of critical skills shortage from corporate sources were not taken at face value and jostled against vociferous accusations that the "shortage" was a myth created by the industry in order to import cheap labor.[8] Industry definitions of skills shortages generally refer to the gap between the estimated demand for workers with a

15

particular skill and the numbers available at that very moment, and never counted the unemployed IT workers who could learn the requisite skills quickly.[9] Employers were in practice so particular about a worker's skill area (or "platform") that, according to Professor Norm Matloff's (1998) congressional testimony, on average only 2 percent of all applications for software jobs succeeded in the United States. In Sydney, big IT placement agents admitted to me that rejection rates of over 90 percent were common in the applications they processed, and I personally came across a case where an agent required applicants to have two to three years' experience in a technology that had been in use for less than one year! Furthermore, most of the reported forecasts of skills shortages were produced by large labor-placement agents, based on surveys of employers. In Australia, the largest IT labor-placement agents such as Candle Australia Ltd., ICON Recruitment Pty. Ltd., and Morgan and Banks Technology all carried out periodical surveys on market trends and publicized part of the results regularly through press conferences and releases by specially hired media consultants. Suffice it to say that IT placement agents themselves favored a tight labor market and were in a position to create a perception of an urgent and perpetual need for IT professionals.

The IT industry argued, however, that skills shortages should not be addressed in terms of simple demand-supply equations but had to be "*over-addressed.*" As a president of the Australian Computer Society (ACS) suggested, only when an excess labor pool was created would more multinationals be persuaded to locate their operations in Australia to tap the immediately available and affordable labor force.[10] In other words, an increase in labor supply would create the necessary increase in demand. Official speeches and documents in Australia had credited great successes to this new official approach.[11] The policy of over-addressing also fits well with the overall development strategy of the IT industry. Ironically, though software solutions are meant to rationalize, standardize, and automate business operations, the process of software development itself has remained snagged by uncertainties (Eischen 2000, 31). To deal with this, the industry over the last thirty years has essentially adopted extensive—rather than intensive—growth strategies; that is, through increases in software manpower rather than increases in productivity or quality through "rationalization." Thus, "even the most celebrated success stories . . . rely on large amounts of unpaid labor to overcome the fundamental inefficiencies and bottlenecks of the software process itself" (Eischen 2000, 33). This development strategy inevitably renders the IT industry highly sensitive to skilled-labor shortages. For example, in response to a proposed bill to reduce the number of H-1B visas in 1995, Microsoft's CEO Bill Gates threatened: "If you want to prevent companies like ours from doing work in the United States, this [bill] is a masterpiece."[12]

In sum, whether or not there was a real gap between IT labor demand and supply is less important; what matters more is employers' desire for an ever enlarging labor supply to maintain the momentum in their expansion. Unlike a real shortage, a *virtual* shortage like this can never be balanced out, as more supply is likely to create more shortage. Thus, the coexistence of a skills shortage and a significant level of professional unemployment can be a long-term feature of the New Economy, a feature epitomized by the routine practice of benching workers in body shops even as more are being hunted.[13]

A virtual labor shortage could not be balanced out, but could disappear abruptly, due to its close links to the speculative movement of capital. This was clearly manifested in the dot-com saga of 2001. Dot-com companies created intangible products: Web sites and Internet-based services to attract large enough numbers of visitors to the site, which they could use as a leverage to sell the site. Tirelessly repeated success stories include Hotmail.com, set up by a U.S.-based Indian software engineer, Mr. Sabeer Bhatia, which was sold to Microsoft for USD 400 million in 1998, and Junglee.com, an Internet browser founded by four Indians, also based in the United States, acquired by Amazon.com for USD 180 million in the same year. Dot-com start-ups were usually sustained entirely by venture capital funds for between six months to two years, when they were either bought up or closed down. In the market meltdown of 2001, venture capitalists panicked and suspended most investment activity[14] and as a result 330 dot-com companies closed down during the first half of the year.[15] No less dramatic were the tens of thousands thrown out of jobs as, due to the hit-and-miss nature of dot-com endeavors, start-ups relied on a huge pool of temporary contract workers who could be laid off summarily in a slowdown. One estimate suggested that 35,000 dot-com workers lost jobs between January and March 2001 in the United States alone.[16] More strikingly, according to Challenger, Gray, and Christmas, a placement firm in Chicago, dot-com companies announced 98,522 layoffs in November 2001, more than doubling the 41,515 firings in 2000.[17]

Gangadharam Atturu, a young Telugu IT professional whom I met on the plane from Kuala Lumpur to Hyderabad, was in San Francisco in early 2001 and recalled how he had started performing *puja* on Friday and Monday mornings, the two days on which IT firms usually announced layoffs. Nothing, however, could sufficiently fortify him for the nerve-racking moment of approaching his desk on those mornings to possibly find the customary "go home" note, or switching on the work computer and finding an e-mail to the same effect. More senior employees might be called in for a talk with the manager, to explain the need for "rationalizing," "right-sizing," and "re-engineering" the company—the words "layoff" or "fire" never coming up. Given the scale of the downsizing in the industry,

17

some placement agents came up with a new business of "outplacement" packages to help firms redeploy their employees to jobs elsewhere, including in other countries; countless laid-off workers had to move on to a new country to search for jobs, or go home—indeed, when the dot-boom became a dot-tomb, "B2C" meant "back to Chennai" and "B2B" stood for "back to Bangalore."

IT workers' global mobility was also facilitated by the fact that standards in the IT profession are set by big global corporations and are thus free from national qualifications and accreditation procedures that have posed a major obstacle to the international mobility of other professionals.[18] Certifications issued by industry players (e.g., Sun Certified Java Programmer, Microsoft Certified System Engineer, Cisco Certified Networking Associate) formed the common criteria for measuring IT skills and were more valued than even a postgraduate university degree.[19] Among my Indian informants, these certificates often took pride of place, securely framed and prominently attached to the wall or carried about at the ready. Venkate, who used to work in a life insurance company in Hyderabad before moving to Sydney in late 1999, had his credit-card-sized Oracle Level 8 certificate in his wallet, facing a portrait of Shri Sai Baba of Shirdi[20] and a photo of his family. Being internationally standardized by global corporations, an informant in New Delhi quipped, meant that IT professionals "can go anywhere in the world—just like e-mails"!

Thus, as a result of the movement of globalized capital and the development strategies adopted by the IT industry, temporary and multiple labor mobility became the norm in the industry. For example, whereas the numbers of "computing professionals" entering Australia as permanent immigrants remained the same between 1997–98 (573) and 1998–99 (574), those arriving as temporary migrants in 1997–98 numbered 3,200 (DCITA et al., 1998) and nearly 4,000 in the year after (Birrell 2000, 80)—or six times the number sponsored for permanent entry (Ruddock 1999, 10). To meet IT companies' concerns for an urgent labor demand, the Australian government created a "contingency reserve" with an extra 5,000 quotas in 1999 under the Employment Nomination program that encouraged employers to sponsor professionals to migrate to Australia as permanent residents, but despite the outcry about skills shortages, the reserve was hardly touched as employers invariably preferred temporary workers (Birrell 2000, 80–81). With the ever changing technologies and job-market situation, a high level of mobility also became crucial for workers' career advancement. Panika from West Bengal in eastern India, working as a Web-development consultant in Sydney, was all set to move to Canada in two months through a body shop when I met him. No different from my Telugu informants, he explained his decision to keep moving as natural:

It is my fundamental idea to go further. . . . You have to do surveys often [by browsing Web sites] to see where you are. You have to follow the latest technologies, and know what salaries they are getting. The way we think is: what's the position I will have in two years? In three years time I may go to America. IT people are never happy with one place. . . . The plan is always open. The train can go anywhere.

Whatever the causes and motivations, mobility of such a high frequency and large scale needed to be facilitated and organized by specialized institutes.

War for IT Talent and Wall of Regulation

If Bill Gates were to arrive in Australia tomorrow to establish a new enterprise, he would be able to get as many IT staff as he wanted on long-term temporary visas.
—Philip Ruddock, Australian minister for immigration (1999, 10)

The continuing high levels of transnational mobility of IT professionals certainly stands as an example of disembedding capital interests triumphing over those of nation-state-embedded regulation, because this global mobility of bodies would only have been assured through deliberate changes in state policy. Australia provides a typical case for how the restructuring of the global economy has compelled the state to facilitate the mobility of IT skills. Australia had always relied on foreign capital and foreign labor for its economic growth. The strategy of importing foreign capital for industrialization and using the trade surplus from raw material and primary product exports to pay for this, however, became unsustainable by the end of the 1970s (see Pinkstone 1992), and as a result Australia experienced a series of trade deficits and an unusually high current account deficit throughout the 1980s. Following the Reagan-Thatcherism prevailing at that time, Australia deregulated its economy; the Australian dollar became freely convertible in 1983, and in 1985, sixteen foreign banks were permitted to enter Australia (Meredith and Dyster 1999, 269), which led to a rapid expansion of the finance sector. This opened the door for one of the earliest streams of migrant IT professionals from India as large banks set out to recruit IT professionals for their expanding operations. St. George's Bank of Australia, for example, sent recruitment teams to Hong Kong, India, Pakistan, and the Philippines in 1987, and the team to India recruited fifty workers, all from Tata Consultancy Services.[21] Along with deregulation, Australia set out to transform itself from a resource-rich "lucky country" to a "clever country" with vibrant high-tech and

19

service industries. The New South Wales state government, for example, planned to make Sydney a world-class finance center, for which goal an advanced IT sector was thought to be crucial. The Australian government's slogan "Swim with IT or sink without it" underlined the sense of immediacy about the urgent quest for IT skills across society.

An important part of the change in development strategy was the "rationalization" of immigration policies, that is, the tendency for the calculation of economic costs and benefits to take precedence over other considerations, including previously pledged social obligations. This was particularly evident after the Liberal Party took office in 1996. The John Howard government increased the proportion of skilled permanent immigrants from 29 percent in 1995–96 to 53 percent in 2000–2001—mainly at the cost of the long-standing family-reunion program (DIMA 2000). In justifying this policy shift, the immigration minister asserted that Australia would be some AUD 95 per person better off as a result of having more skilled migrants,[22] that whereas every 1,000 skilled migrants added AUD 35 million to the budget over five years, every 1,000 unskilled family migrants would cost Australia AUD 1.5 million, and, further, that every 1,000 refugees would cost AUD 21.5 million.[23] According to the Points Test System and Migration Occupation in Demand List of the Department of Immigration and Multicultural Affairs (DIMA), IT professionals were classified under a Tier One occupation, scoring maximum points (Birrell 2000, 80). In 1999 the Australian government simplified the assessment procedures for employer-sponsored IT migrants as well as for permanent residency for overseas students with IT degrees, expanded and liberalized access to the DIMA Skills Matching Database, and conducted an awareness campaign, specifically targeting IT employers, to publicize DIMA's various IT-friendly programs (Alston et al., 1999).

The most important measure of rationalization was perhaps the streamlining of the long-term temporary migration scheme for professionals, namely the 457 visa scheme, in 1996. The new scheme exempted those sponsored by Australian companies from any market testing or skills assessment requirements, and attempted to be as responsive as possible to the demand from industry (DCITA et al., 1998). The "[l]ong-term skilled temporary migration has become a core element of Australia's immigration program," declared the Australian immigration minister (Ruddock 1999, 6), who went on to suggest that the reform would make the scheme "rated amongst the most effective arrangements in the world for attracting what is a highly mobile and lucrative workforce . . . [and] will become the touchstone for migration's international future" (Ruddock 1999, 10). The proportion of all migrants entering Australia on 457 visas increased from the initial 4.4 percent in 1991 to almost 20 percent in 1996, and to over 50 percent in 1999–2000.[24] Under this scheme IT professionals, once again, were given "immediate processing priority" (Alston 2001).

This rationalization of immigration policy was far from unique to Australia. All the major developed countries significantly relaxed migration regulation for skilled labor throughout the 1990s (United Nations International Economic and Social Affairs 1992; OECD 2001; 2002). The U.S. Congress, for example, increased the annual quota of the H-1B visas from 65,000 to 115,000 for the fiscal years 1999 and 2000; and up again to 195,000 for the following three years. Furthermore, government and semi-government bodies reached out aggressively to recruit overseas IT workers, particularly from India. For example, the New Zealand Immigration Services, Trade New Zealand, and the Information Technology Association of New Zealand jointly organized a delegation to India in October 2000 to recruit two hundred IT experts; the Immigration Services even set up a special unit in India to assist IT professionals to migrate to New Zealand.[25] As the minister for communications of Singapore, whose government had even proposed a visa-free travel scheme within Asia for IT professionals,[26] aptly summed up the situation: "We are in a global war for [IT] talent."[27]

Eager as the states were to embrace global capital and foreign talent, none of them has loosened up immigration or labor regulation altogether. Australia's 457 visa scheme, for example, obliged companies to assume financial responsibility for the foreign workers they sponsored so as to limit the social and financial costs borne by the Australian community. The sponsor had to bear liability for up to AUD 2,000 per month for each worker, which included specific allocations for various living expenses like AUD 1,000 for housing rent, AUD 200 for telephone bills, and so on—in other words, if a 457 visa holder defaulted on his rent, the sponsor would be held liable for up to, in this case, AUD 1,000 for each month. So a firm sponsoring ten workers for two years faced possible liabilities amounting to AUD 480,000, not to mention the human resources and time spent on sorting out such problems, which an IT firm in a volatile market could hardly afford to waste. More troublesome than this, employers, particularly those in the state of New South Wales, are subject to a complex award system according to which an employment relationship is judged on substance rather than form. Thus, a court may deem the status of a long-term temporary contract job to be one of permanent employment, notwithstanding the existence of a contract (Nolan 2001, 4). Because of this, I was told, an employer often could not terminate a contract with a temporary worker.

In the United States, a company intending to sponsor an H-1B visa applicant in the late 1990s had to go through a procedure involving at least

21

four government agencies, namely the local employment service, the Department of Labor, the State Department, and the Immigration and Naturalization Services. It sometimes took four months just for the INS to process an H-1B visa.[28] From the mid-1990s, Congress required that all H-1B employers should file a "labor condition application" with the Labor Department to protect American workers from being displaced. Given that the regulations for this procedure ran to 150 pages,[29] the outpouring complaints that this imposed "significant constraints on the human resource strategies of companies"[30] were to be expected. Furthermore, compliance with various provisions under the labor laws, from conditions of entry to payment, was fraught with difficulty for sponsors in increasingly uncertain market situations. In early 2000, forty Indian IT professionals were arrested and handcuffed by INS officers at Randolph Air Force Base, Texas, for working outside of the site designated in their migration documents—a violation of the H-1B rule that tied temporary workers to a specific employer, field of activity, and geographical area. This triggered strenuous criticism of the regulations by both IT companies and workers who argued that by definition the IT industry knew no geographic boundaries (see posts at www.usvisanews.com in February 2000).

The discrepancy between the fixed term of a temporary work visa and the unpredictable, usually much shorter, duration of IT project-based contracts made it unfeasible for companies to be responsible for a migrant worker throughout the visa period. While the Australian 457 visa is valid for one to four years, most IT projects last only six months. Nor could companies realistically be expected to go through the entire sponsorship procedure for every worker they employed on ever short-term projects. Moreover, the intensification of alliance and cooperation between companies made it difficult to even identify who should bear responsibility for sponsoring workers on a joint project.

These frictions between state regulation and the fast-changing requirements of technology and maximizing capital investment created the special niche for placement agents in general and body shops in particular.[31] Placement agents and body shops were essential for an institutionalized flexible labor market, not only for matching up workers and companies, but more importantly for surmounting institutional obstacles. For example, with a placement agent taking on the function of signing contracts with workers, the eventual employer, now as the agent's client, had no direct legal relationship with the worker placed with them, and no onerous liabilities. By specializing in the tedious recruitment and migration procedures, placement agents and body shops could attend to likely problems much more efficiently, for example by maneuvering between different types of visas (tourist, short-term business visitor, student) to get workers into a country quickly, and convert their status with a working visa afterward.

Uday, a twenty-nine-year-old IT worker from Andhra Pradesh and my key informant, contacted a body-shop-style company in Singapore in 1998 when he had only fourteen hours left on his tourist visa. Within two hours the company arranged for another short-term-visit visa, the very next day they sent him to work, and within a few days had sorted out his work permit. In Australia too, it was fairly common that body shops brought workers in on 456 visa and then converted it to 457. The 456 visa was granted for the purposes of official meetings, training or equipment installation, for a period of up to three months (normally referred to as "business visa"), and it was easier to obtain since it precluded employment. Precisely because of the friction between the state and corporate interest, the 2001 market crash in a sense made body shopping more, rather than less, important for labor management of the global IT industry: body shopping enabled swift and large-scale layoffs without negative impact on the domestic society of the destination country.

Chapter 2
Producing "IT People" in Andhra

Just after I landed in Hyderabad in June 2001, Gopal, a young IT worker, assured me that my field research could be completed within a month as I would easily identify and round up informants: "Anyone in the street who puts his shirt inside his trousers [in the Western style] is an IT professional." Vinnie, who had first gone to Australia as a student and was running a body-shopping business in Sydney, first heard the term "Y2K" in 1998 on a call to his mother in Hyderabad. Vinnie had to ask his mother, who had only six years of education and spoke no English, what she meant by "Y2K"! (Only after 1999 did the "Year 2000 Problem" become known in Australia by the popular American acronym.)

In the "epicenter of India's cyberquake," as Hyderabad was described in the public media and official documents, IT was part of political chant. Soon after becoming the chief minister of Andhra Pradesh in 1995, Mr. Chandrababu Naidu identified developing the IT sector as a top priority. Calling himself the "CEO" of Andhra Pradesh, he set up India's first Department of IT (dubbed "DoIT") and, in a bid to transform the city into a world center of software development, designated northeastern Hyderabad as an IT-based special-development zone—the snappily named Cyberabad. In 1996 Naidu secured an appointment with Bill Gates,[1] and in 2000, by stressing a booming IT industry, Naidu successfully persuaded the former U.S. president Bill Clinton to visit Hyderabad during his landmark visit to India. Fashioning himself as the state's quintessential "ABC"—Andhra's Babu Computer—Chandrababu Naidu was voted by *Profit*, a magazine published by Oracle Corporation, as one of the "seven doing wonders in the world" for 2001.[2] All this media and official hype, however, was somewhat at odds with the reality on the ground. Andhra Pradesh's share of India's software exports actually dropped from 9 percent in 1994–95 to 5.73 percent in 1998–99,[3] and the materialized foreign investments in the IT sector in 2001 were estimated to be a very modest INR 500–600 lakhs (around USD 15 million) (Software Technology Parks of India, Hyderabad, 2000).

Notwithstanding the noise and heat generated by state government's political maneuvering, there was good reason for the contagious IT mania in Andhra Pradesh, which was not so much about the local IT industry as it was about "*IT jobs*," jobs that guaranteed unbelievably high incomes, and what's more, opportunities to work overseas. Stated more bluntly, by the late 1990s, IT was about celebrating *Dollar Dreams* (the title of the movie of six young Hyderabadi IT workers showing at the time). Ashok, my key informant Uday's brother, who studied and worked in Hyderabad in IT from 1993, had directly witnessed the beginnings of the mania:

> There was almost no IT in AP until the mid-1990s. In 1996, Satyam and Sriven [two major IT firms in Hyderabad] got large [Y2K] projects from the U.S. They came to campus and . . . took people in tens or hundreds! No one had seen anything like this before! A fresh graduate could be paid INR 20,000 [a month]—they are not making money, they are minting money. . . . In late 1997, Microfocus [a software program simulating the IBM mainframe environment] came [to Hyderabad]. Training institutes came out like anything. Then IT really became a big thing in the city.

The consequent growth of the private sector IT training business in India was truly spectacular, with revenues somersaulting from INR 33 crores in 1994–95 to INR 125 crores in 1998–99 (Nasscom 2000), and to INR 2,150 crores in 2000–2001.[4] Central to the IT mania in India, therefore, has been the production not of IT products but of "IT people," who in the everyday life of Hyderabad, as elsewhere in India, came to denote a special *social* category. A fresh IT graduate could in 1999 get INR 10,000 a month in a decent company—almost the highest salary of a senior non-IT engineer in a big company. Casually talking to middle-class Hyderabadis easily summoned up stories of IT people who went to the United States and changed the lives of whole families—or a recounting of "IT folklore" as a Telugu student informant aptly put it. Chawdary, a middle-aged public-relations manager for an IT firm in Hyderabad, commented on these young nonresident Indian (NRI) IT people working in the United States: "A *boy*, twenty-two or twenty-three years old, can put INR 40,000 to 50,000 every month in his NRI account[5] for his family! In dollars! For rich families it used to be rupees make rupees, now rupees make dollars. *Is there any business better than this?*"

It is routinely pointed out that India's successes in IT are due to the industry's nearly exclusive reliance on cheap labor resources, and favored by the limited requirements for financial capital. What such assessments neglect to acknowledge, however, is that the *production* of the IT labor force itself requires the mobilization of large and long-term investments. I call it "production" because the process comprises collective efforts involving systematic and institutional arrangements and is an integral part of the overall socioeconomic life of the local society. This chapter delves into the key domains of the production of IT labor in Andhra Pradesh, namely higher education, specifically private colleges and training institutes; caste; family networks; and particularly, the institution of dowry.

"D-Shops" and "T-Shops"

The relatively well-developed tertiary education and research system in India is often credited for the high presence of Indians in the global IT industry. In this connection, Andhra Pradesh hosts ten nationally funded research centers (e.g., the Center for Material for Electronic Theory) and seven national military research institutes (e.g., the Defense Electronics Research Laboratory). Substantial numbers from the large pool of professionals created by such institutions later became pioneers of IT development in India. On the higher-education front, the number of universities in India increased from 37 in 1950 to 129 in 1975, and engineering colleges increased from 58 to 179 (Krishna and Khardria 1997, 351, table 2). The seven elite Indian Institutes of Technology (IIT) set up since 1950 modeled on the Massachusetts Institute of Technology (MIT) have established a formidable reputation as "the biggest Indian brand after the Taj Mahal"[6] as well as becoming de facto incubators for West-bound emigration: by the 1990s, IIT graduates accounted for 40 percent of all engineering graduates migrating to the United States (Krishna and Khadria 1997, 363). One media story described the exodus of graduates from IIT Chennai as such an established phenomenon that "the local campus postman and bank clerk provide unsolicited advice on the best U.S. schools to attend. . . . When acceptance letters arrive, the postman waits outside the student's door for a tip—a large one if it's from a highly regarded university such as Stanford."[7]

But with the rapid expansion of opportunities generated by global body shopping, it is private-sector institutes and not public universities or elite institutions that are churning out the majority of today's hypermobile IT people. Between 1995 and 2000, seventy-five private engineering colleges were set up in Andhra Pradesh—compared to twenty-six (government and private) over the sixty years from 1929 to 1989 (SCHE 2001a), and eight between 1990 and 1995 (SCHE 2001b and 2001c)—almost all delivering IT education. By the end of 2000, Andhra Pradesh had ninety-six engineering colleges offering computer-related courses, 469 colleges (excluding the engineering colleges) offering bachelor of computer application (BCA) degree courses, and 161 offering master of computer application (MCA) (SCHE 2001d). In total, nearly 100,000 students in the state were enrolled in IT or IT-related courses in the year 2000–2001 (SCHE 2001d).[8] Education at these colleges was strongly emigration oriented, and their IT curriculum was often pulled entirely from U.S. textbooks. As part of their "placement services," colleges introduced students to international education agents who recruited students for universities overseas, particularly in Australia, Canada, the United Kingdom, and, increasingly, Germany; some college staff were themselves agents for universities in Australia.[9]

Private colleges absorbed considerable private resources. Government regulations on facilities/equipments and qualified staffing required a minimum investment of INR 5 crores to set up a private engineering college and INR 2–3 crores for a normal degree college. During the year 2000–2001, five new private colleges were set up in Andhra Pradesh (CTE 2001), and investors poured around INR 5–6 crores (of the total investments of INR 15–20 crores) into IT subjects.[10] By far the largest investments in IT education, however, were the resources mobilized to pay for the tuition fees, which in 2000–2001 would have been around INR 200–240 crores.[11]

Along with the private colleges, private technical and technology training institutes that could not grant degrees mushroomed. While some private colleges were called "D-shops" since they notoriously sold degrees without imparting any practical knowledge, private IT training institutes were referred to as "T-shops," and supposedly sold technical skills. Because IT companies expected new employees to already have hands-on experience, it became a de facto compulsory requirement for IT students to take courses in T-shops while pursuing their degrees, whether at private colleges or public universities. One T-shop in Hyderabad therefore advertised its courses as "specially tailored for OU [Osmania University][12] MCA students!" My interviews with students and staff at the Indira Gandhi National Open University and the Sultanululoom College of Administration, both in Hyderabad, confirmed that about 70 percent of their IT students in 2000–2001 were simultaneously attending technical courses along with their college courses (financial difficulty was given as the main reason that the remaining 30 percent were not doing so). Since training institutes required much less investment than colleges and tended to be much smaller in size, the number of T-shops in Andhra Pradesh far outstripped private colleges. Within a one-kilometer radius of the Hyderabad Urban Development Authority (HUDA) building on Ameerpet Road, northern Hyderabad, there were perhaps 2,000 to 3,000 students attending courses every day at some 120 or so training institutes in early 2001.[13] Uday recalled that when he had just moved to Hyderabad from his home-town Vishakhapatnam in 1998, an industrial town in northeastern Andhra Pradesh that had yet to be engulfed by the IT fervor at that time, he hit four or five IT training institutes before finding a barber to have a haircut.

Private IT training institutes set it as their singular mission to produce IT workers immediately employable in the West, though institutes of different sizes adopted different strategies in doing so. The National Institute of Information Technology (NIIT), for example, stood out as the biggest IT training institute in Asia, with 8,000 staff and 600,000 students on roll in 2,228 learning centers in forty countries in early 2001 (NIIT 2001). NIIT's business success was first of all based on its global connections.[14] In 1996 NIIT prepared around 18,000 students to take the Microsoft Certified System Engineers (MCSD) examinations, and about 2,000 had passed within six months. Microsoft was so delighted to see Microsoft engineers being produced so efficiently on such a mass scale—having a large group of technicians familiar with the company's software was key to maintaining market share—that it designated NIIT as its "premium education and training partner for India, greater China and Sri Lanka." With this status, NIIT could obtain software programs from Microsoft one month before their public release, and therefore gained the critical "first mover's" advantage in the IT training industry: NIIT had won Microsoft's best training company award every year from 1998 to 2001. Besides Microsoft, NIIT had also tied up with Oracle, Red Hat, Computer Association, and CITRIX (a global leader in system software for server-based computing), and all these companies released their latest technologies in India through NIIT; in return, NIIT was obliged to produce a certain number of users of these technologies within given time frames.

The content of NIIT's training programs was closely tuned to the latest technologies and global-market trends through its research and development centers in New Delhi and Atlanta. A curricular emphasis on cross-cultural communication to prepare students to work abroad was evident in its four-year flagship program, iGNIIT, which exclusively targeted college students and included forty to forty-five hours per semester on effective communication and personality development (40 percent of students on the program reportedly worked overseas on completion). In addition, "human power placement" courses taught skills useful for overseas job interviews and presentations, while part of the "career development" courses was about workplace cultures in the United States and Europe. Similarly, Tata Infotech also provided comprehensive "soft skills" training in its G-Tech program, ranging from "customer care and orientation" to "behavioral skills," aiming to produce "complete infotech persons" imbued with certain values, attitudes, and even gestures, and not mere "engineers" doing back-end work.[15]

For the continually sprouting medium- and small-sized T-shops, however, the rush for global connections brought a quite different fate. At least 15 out of the 120 or so institutes in Ameerpet completely closed down during the slump in the first half of 2001, and another 20 that still had their names on the door had stopped operating.[16] Medium-sized institutes once well known in Hyderabad, such as Netgac, Java Point, Eureck, and Win Tech, also shut down in 2001.[17] Rajashekhar Reddy, the founder and principal of I-Logic, a small training institute that was still in business in 2001, attributed the dynamics of these short life cycles to "U.S. rule":

> In Hyderabad, you must offer the course immediately when students ask for it. Hyderabad is completely ruled by the U.S. If you have five friends in the U.S., and three say Java is good, you go to Java immediately. Then everyone goes to Java. The next day, people in America say Oracle is good, everyone goes to Oracle. . . . Now [in the slowdown] the U.S. can't give any suggestion. The only suggestion is not to do Java. This is why so many institutes have closed down.

Most small institutes' response to the fallout wrought by "U.S. rule," paradoxically, was to follow trends in the United States even more closely. Information on the U.S. market was gleaned from students who came around to "order" specific courses that their friends or body shops promised would get them openings there. Some smaller institutes took the cue from big ones, even sending their staff to be trained there in the latest courses during the day, and then teach these to students in the evening! Others kept a close watch on trends set by the major world IT companies. In the case of I-Logic, Rajashekhar Reddy looked to Oracle: "I follow everything Oracle does. When the Oracle 8i came out, I studied it thoroughly to know what exactly the advantage is and what is the problem. When Oracle 9i comes, I can immediately know what it is for, what problems it is supposed to solve, then I can teach my students." Training institutes were so eager to be ahead of the technology curve that courses were sometimes offered prematurely. For example, taking the lead from NIIT and Aptech (the second largest IT training institute in India), a large number of institutes launched "C#" and ".Net" courses in early 2001, even before it was clear that any company in the world would have real projects requiring these skills anytime soon.[18]

But, no matter how differently they survived or anticipated the changes in global markets, all training institutes ultimately relied on the tuition fees from students for profit. NIIT's global expansion of business, for example, was also developed through sophisticated local strategies for recruiting students, mainly relying on franchises. Compared to branches or subsidiaries, franchises have deeper local roots and can more effectively penetrate the local market. At one of NIIT's franchises in Tanuku town, West Godavari district in eastern Andhra Pradesh, two full-time executives specialized in "village marketing." Students at the Tanuku institute were asked for information about other college students in their home villages; the two executives followed up by tracking down those students' families and persuading them to send their sons to NIIT, sometimes with students from the institute in tow to provide convincing testimony. Over the month of June 2001, the two executives visited thirty-four families and recruited eight students. The Tanuku institute also offered incentives to its students to rope in their friends—a gift voucher worth INR 500 and two NIIT lottery tickets with a chance to win a car or a computer for each student brought in.[19] NIIT's successful approach in business—combining global connections with local strategies—was no small reflection of the significant change in the scales of the local resources being channeled into producing globally mobile professionals in India, where the countryside became the major source of surplus value.

"Have Lands in Andhra, Have a House in Hyderabad, and Have a Job in America"

Compared to earlier streams of migrant professionals, most notably of those in medicine, the majority of the IT people moving out of India were clearly distinguished by their rural town, including agricultural, background. While as recently as the 1970s most migrant professionals came from elite families in metropolitan areas, now considerable numbers of well-off families in the countryside produced software professionals through investments in private education.[20] However, given that the majority of rural families could never afford a private education, this social background did not make the IT people any more representative of Indian society than metropolitan classes.[21] A survey in 2000 found that 81.8 percent of software professionals were from forward castes, and only 9 percent were from the backward castes comprising about 52 percent of India's population (Vijayabaskar et al., 2001).[22]

Nor did the rural background of most IT professionals mean a more even distribution in geographical origin, which calls attention to the specific local setting of the production of IT people. It was commonly held that 70 percent of Telugu IT professionals originated from coastal Andhra, one of the three regions composing Andhra Pradesh (coastal Andhra, hinterland Telangana, and Rayalaseema between them), and further, that 80 percent of these 70 percent hailed from just four districts located in the Godavari and Krishna river deltas—East Godavari, West Godavari, Krishna, and Guntur (see figure 1). (These figures were circulated among Telugu IT professionals in the United States and, consequently, widely cited by those in Andhra Pradesh.) Coastal Andhra in general and the four districts in particular are among the agriculturally richest areas in India. West Godavari, for instance, had 5.3 percent of the state population (DES 1999, 3, table 1.3) but 10.6 percent of the state's rice fields in 1998 (DES 1999, 71, table 4.9) and a per-acre food grain output that was 60 percent higher than the state average (DES 1999, 71, table 4.9). However, despite the surplus produced, and the fact that nearly 90 percent of the district's population still worked in agriculture (DES 1999, table 1.8), little resource went toward industrialization of this sector. Instead, a large portion of agricultural surplus was diverted into higher education: in the nine villages that I investigated (seven in coastal Andhra), higher education absorbed on average about 25 percent of the agriculture surplus.[23] State-wise, 80 percent of the overall investments in private colleges were estimated to originate from agriculture surplus, deriving mainly from coastal Andhra.[24] Most private colleges were in fact joint ventures between large landowners, educators such as retired principals, and local politicians. The impetus for why and how such a significant amount of surplus was diverted to higher education sprang from the caste relations and social-mobility mechanisms in the local society.

The majority of the IT professionals from Andhra Pradesh came from the two most dominant forward castes, the Kamma and Reddy, with a smaller group of Rajus. Both Kammas and Reddys were originally cultivator castes and are still closely linked to the rural agriculture society. The Reddys had close links with the Congress Party and had dominated state politics in Andhra Pradesh after Independence in 1947 until 1984, when the legendary movie star from the Kamma community, T. N. Rama Rao, became the state chief minister. The Reddy population is larger, more heterogeneous in economic status, and more geographically dispersed, while the much smaller Kamma community is concentrated in the four coastal districts, particularly in Guntur, and is regarded as the richest group in Andhra Pradesh. (As much as 50 percent of the large businesses in Hyderabad were said to be owned by Kammas.)

The rise to dominance of Kamma and Reddy in the last century owes much to education. In the 1920s, both Kammas and Reddys were leading forces in the anti-Brahmin movement in Andhra Pradesh,[25] and to circumvent the Brahmin monopoly of education, they set up in town Western-style schools, ostensibly open to all castes though the majority of students were from the two communities. With Independence, and the attending creation of a large number of secure public-sector jobs, higher education—once perceived as an elite Brahmin and colonial British preserve—became a valuable asset that held out much greater promise for all middle- and upper-caste communities. In the 1950s, Kammas set up their own *sanghas* (associations) in Andhra Pradesh to provide its members scholarships for higher studies. The Green Revolution beginning in the 1970s and the introduction of Chinese TN1 rice seed in the 1980s significantly increased agriculture surpluses in coastal Andhra. This consolidated the dominant position of the Kammas and Reddys as rich farmers and landlords and fostered the emergence of a "rural middle class"[26] mainly comprising formerly poor Reddys, as well as Kapurs (a large middle caste in Andhra Pradesh). These families too saw education as a means to fortify their newly achieved status. Quite commonly, well-off families in Andhra send their children off to the neighboring states, particularly Karnataka and Tamil Nadu, where there are more private colleges; every year, over 15,000 students leave, taking with them a minimum of INR 8–10 lakhs to pay for their academic courses.[27]

Partly to obtain a good education for their children, since the mid-1980s large numbers of wealthy farmers bought houses in town to settle their families. This fueled the development of towns like Tanuku and Bhimavaram in West Godavari. While the father moved from village to town, the son aimed to move from town to metropolitan cities and beyond, the West. The model Kamma family strategy was said to be, as a Reddy put it, "to have lands in Andhra, have a house in Hyderabad, and have a job in America." But continuous outflow of local resources left the rural areas underdeveloped. The low level of capital formation in agriculture rendered farmers highly vulnerable to pest infestation and unfavorable weather. According to Dr. M. R. Raju, a physician who worked in the United States for more than thirty years before returning to his native West Godavari for charitable work, the living conditions of the poor in his home village, Peda Amiram, remained basically the same as half a century earlier (see Raju 1999). Chronic underdevelopment in the rural areas thus encouraged families with the means to invest even more in education in order to move out.

A central element to the historical dynamics of caste, social mobility, and education is the institution of dowry. A family with a successful IT son most often attracted attention or jealousy from the neighbors because of the dowry that he would summon. The Kammas in Andhra Pradesh are known for their high dowry rates. Reaffirming Srinivas's (1996; 1983) analyses of dowry as a historically recent institution in southern India,[28] my field interviews and observations in villages in coastal Andhra suggested that dowry had been practiced for less than one hundred years, and in many cases perhaps only thirty.[29] In Gurrapaadiya village in Prakasam district, on the east coast of Andhra Pradesh, I found out that the first dowry marriage occurred only in 1996, when the first highly educated man in the village, a doctor, had demanded a dowry. Mostly, the village had practiced either exchange marriage or marriage arranged before or at birth among relatives who formed a marriage circle and, only occasionally, marriage that carried the burden of bride price.[30] Since 1996, however, dowry had fast become the norm among the educated youth in the village, even between matched cousins. A first-year MBA student from the village whom I met at an English coaching class in Bhimavaram was approached by his paternal uncle with a dowry offer of INR 10 lakhs.

Modernization and global mobility have not, it appears, made local society in Andhra Pradesh any less orthodox when it comes to marriage. Arranged intra-caste (*jati*) marriage remains the norm; those marrying outside of their castes are considered suspicious and unstable, which often jeopardizes their siblings' marriage prospects. Where marriages had been matched within a limited circle of known families, and dowry given less importance, with the unprecedented and expanded scope of the marriage market in cities and towns, and in cases of NRIs and IT people, overseas, matches are quite commonly sought through the newspapers, Internet, and marriage bureaus. Once the marriage market is disembedded from social relationships, sheer economic calculation overwhelms other considerations, and along with traditional caste and horoscope compatibilities, occupation, income, and property are ruthlessly assessed and compared. Also, most parents still feel obliged to marry off their daughters as early as possible. In many places, a family with a twenty-year-old daughter at home is a subject for gossip; having a twenty-five-year-old unmarried daughter is scandalous. A UNICEF study in 2001 found that 80 percent of the females in Andhra Pradesh were married before age eighteen, and 50 percent before they turned fifteen![31] I heard of a Harvard University graduate and another about to pursue a doctorate at Oxford who both happily married nineteen-year-old girls. The preference for young brides made investments in their education less likely, and consequently high dowries more necessary to compensate for this "inferiority." Apart from that, parents prefer brides from village families, and certainly from India, while encouraging educated sons to migrate to the city or the West, which has led to a scenario of marriages matched between a bride from a rich village family and an educated city-based groom, or between an Indian bride and a groom overseas. This preference for upward mobility on the part of the bride's family is worth a high dowry that thus serves as a direct means to transfer surplus value from the rural economy to the urban, and from the local economy to the global.

Producing IT People as a Family Business

While colonialization and modernization precipitated the institutionalization of the modern dowry, the era of globalization and the appearance of transnational IT grooms have bestowed the practice with a revolutionary impetus. In 2000 an IT groom from the Kamma or Reddy caste and working in a big company in Hyderabad expected to receive a dowry of INR 8–20 lakhs, more than double that for a non-IT engineer. Previously, a groom received a high dowry only if he was working in the public sector or the bride's kin were satisfied that he had secured a promising future in a private company. But IT qualifications could by themselves—sometimes without the grooms having proper jobs even—command high dowries. (Incidentally, dowries in 1998–2000 commonly included IT stocks.) If the IT groom was based in the United States, the dowry could go up to USD 120,000. This doubled the worth of an Australia-based groom, partly because the value of the U.S. dollar was equivalent to double the Australian. The high conversion rate for pound sterling relative to other currencies, however, did not favor grooms working in the United Kingdom, because it was believed that the net savings possible in the United Kingdom was only two-thirds of that in the United States. Most families that I visited in urban coastal Andhra had intimate knowledge of the incomes and expenditures of Indian IT workers in the United States, and some were conversant with those in the United Kingdom and Australia as well.

Despite their high incomes, the Indian IT men working overseas were no less indifferent to inducements of dowry—and sometimes even despite their already married status! Quite a few informants in Sydney voluntarily mentioned dowry as the key criterion when discussing the marriage proposals they received and how they processed the bids. In Hyderabad, over a few years at the end of the 1990s, Professor T. R. Sudarshan Rao, a former principal of a law college, had been approached for legal advice on no less than twenty cases of bigamy committed by NRIs. An ongoing case in mid-2001 involved a man in Silicon Valley who pulled off a marriage in India after having married in the United States. These men had partly been pressured to marry the bride chosen by the parents, and, suggested my informants, not surprisingly were partly attracted by the dowry: "Who doesn't like money?" On learning the truth, some brides' families were determined to seek legal redress while others decided to keep up appearances, having all too widely publicized the fact of "a U.S. son-in-law"!

For the groom's parents, high dowry was viewed as a direct reimbursement for their investments in the son's education. It was common at marriage negotiations to present the bride's kin with a detailed breakdown of the costs of the groom's private education over the years and demand dowry accordingly. Although it was claimed that "good families" demanded dowry for the young couple, in most cases the dowry was presented to the parents who then decided how it would be portioned. Some recent changes in the composition of dowry have also benefited the parents more. For many lower-class families across Andhra Pradesh, dowry consists mainly of gold, cash, furniture, and utensils (which is the norm in northern India), but for wealthy families, expensive articles and real estate become increasingly popular. In one case, an IT groom from the Padmashali community[32] received a dowry of INR 15 lakhs, including a three-bedroom apartment in Hyderabad worth INR 5 lakhs and a car worth INR 5 lakhs. The apartment and the car were used by the groom's parents since the young couple went to the United States. This was an ideal reconciliation between the rhetoric about dowry as a conjugal fund and the calculation of dowry as returns to the investment in the next generation: the groom's parents could enjoy the dowry without taking it over, unlike in the case of cash or gold.

In small towns and villages, dowry was sometimes paid in a manner of a "forward selling" or "futures market." A bride's father may offer to pay an IT boy's college fees or sponsor his emigration on the condition of a subsequent marriage. An engagement ceremony would publicize the deal in the community or a civil marriage would be registered before the groom proceeded overseas. Then, while the bride started the application process for her dependant's visa, the groom was supposed to secure a job and find a place to settle before returning for the religious wedding ceremony and then taking the bride back out with him. Citing this practice, some men now used enrolling on IT courses or wanting to go abroad as an excuse for demanding additional dowry after marriage.[33] One informant in Hyderabad lamented that he would have gone to the United States long before had his in-laws been more generous.

Dowry as sponsorship did not, however, always work in the groom's favor. Vijay Naidu, a civil engineer by qualification, received a dowry worth INR 7.1 lakhs when he married in 1998. His father-in-law and elder brother-in-law urged him to switch to IT and go overseas. Vijay quit his job to take full-time IT courses and his wife brought in INR 50,000 from her brothers to support him, but after two years he still could not get a placement abroad. As a result, he told me, he was suffering from "chronic mental torture": "Every time I visit them, all the aunts, uncles, brothers-in-law ask: Why are you still here? When will you go? I have to lie; my wife has to lie." He was glad for the slowdown in the United States, which gave him a legitimate reason for being at home and carrying on with IT training. In another case, a bride from Tanuku paid INR 25 lakhs to support the groom's studies in the United States. But, when she joined him after marriage, she could not cope with American English and the life without servants, and did not see why she should have to adapt on the grounds that she should not have to suffer after having paid so much dowry—without which the man could not have gone to his future in the United States. She returned to India and it was not clear how the husband dealt with this. An even worse situation resulted when a groom granted a high dowry lost his job overseas. Narendra, a Telugu IT worker, went to Sydney in early 2001 with the airfare paid for by the would-be in-laws, but could not find a job for five months. The in-laws urged his family to finalize the marriage and his family pushed him to do so too. Narendra could not tell them the truth; nor could he go ahead with the marriage without a job and without the money to buy gifts from Australia.

IT people not only attracted high dowry but, at least until the slow-down, their high incomes and remittances also increased their families' capacity to pay high dowries for the sisters. Overseas remittances were usually sent back unconditionally in the first year or so; after that, instructions might be given or explanations required for how the money was used. But securing good grooms for sisters was always a top priority. Apart from the moral obligation for brothers to help the father with their sisters' dowries, it was also in the interests of IT men to do so. In most parts of Andhra Pradesh the convention was to marry off the daughters before taking in daughters-in-law. The dowry offered indicated a family's social status, on which the family could, in turn, base their claims for the son's dowry. Additionally, marrying off all the daughters removed future burdens on the family's property, which thus allowed potential brides to more accurately calculate the groom's worth as well as a fair dowry offer. From this standpoint a family would not be reluctant to perform grand marriages for the daughters since they could await the dowries for their sons. Dowry giving and receiving were thus inherently linked, centering on the concept of *kutumba gowravam* (family status), and the anxiety to protect, or more often to uplift, the family's status motivated young IT sons to contribute as much as possible to their sisters' dowries.[34]

Since dowry rates for IT grooms were directly tied to their mobility and prospects in the global market, the massive layoff of IT workers world-wide in 2001 dealt a severe blow to their position in the marriage market. My friend Gangadharam Atturu was so alarmed by the market downturn in Los Angeles that he decided to return to Hyderabad in June 2001, just as his father had started bride hunting, and the return spoiled all the plans. The father asked me why George Bush had done this to him. The magazine *Computer Today* reported a 70 percent drop in the dowry rate for IT people in Hyderabad.[35] The response of some IT grooms was to get married before graduating, a strategy explained by the wisdom of the Telugu saying: "Put the buffalo in the water when bargaining for the price." Likewise, a student whose capacity to work can only be guessed at is unlikely to secure a high price, but that is still better than the fate of a rejected buffalo on the shore.

As crucial as dowry in pooling resources to support IT study and emigration was the role of extended family networks. Vijay Naidu, the poor fellow suffering "mental torture," was very proud of his own "real Indian family." After exhausting his dowry, he managed to carry on by relying, first, on his second brother in Chittoor (a townern in south Andhra Pradesh), then on the fourth. When this fourth brother joined a local IT company in Hyderabad in 2000, he urged Vijay, together with wife and daughter, to move from Chittoor to Hyderabad to try his luck. They all lived in the same flat. The brother, who was earning INR 8,000 a month and handing over more than INR 6,000 to Vijay (Vijay's wife was running the house), explained the situation to me:

> My plan is to send my brother to the U.S. first. Then I will be free, I can prepare to go [to the U.S.]. . . . My money won't be enough [to send Vijay abroad]. We have land in the village. We may sell it. . . . No, the other brothers won't say anything [about selling the lands for Vijay]. It's up to my father. If Vijay can go to the U.S., it's good for all of us.

While investigating the social origins of IT people, I noticed that single-caste villages produced proportionately more professionals than the average. For example, Atreyapuram, a Raju village in East Godavari, and Illindila Rapu, a Kamma village in West Godavari, were both well known in the nearby areas for having large numbers of professionals and emigrants. The residents in single-caste villages could often all be traced back to one family, which might have enabled them to mobilize resources more easily. I knew a girl from Gurrapaadiya, a Reddy village in Prakasam district, who was matched at birth to a cousin. Over the ensuing twenty years, the cousin's family remained poor, but, as her family prospered, they paid for the most part of the cousin's education and made him an IT person.

Extended family networks facilitated the systematic and long-term investments and supports necessary for human-resource development that would not be available from other sources. A Muslim family in Guntur district had lost the father early, and the two elder brothers borrowed money to educate the third one as a doctor and the fourth as a computer engineer, who they matched to a U.S.-based bride. Forgoing the offer of a dowry, the family made a deal with the bride's father that the groom should be free to continue sending money back. In another case, Madhu, a Kamma student from Guntur, was invited by his uncle to study in Melbourne. Madhu's father had brought up the uncle and supported his emigration to the Middle East, whence he went on to the United Kingdom, and finally, to Australia. In return for the support he received, the uncle now wanted to enroll Madhu in the postgraduate IT course his own son was in. But family networks are not always sweet and cozy, and bitter comparison among relatives is sometimes more powerful than mutual assistance in pushing people to study IT or to migrate. Uday's brother, Ashok, bought a Suzuki car in 1998, soon after joining the Hyderabad office of Baan, a Dutch software house headquartered in Amsterdam. Immediately, three of his four uncles bought cars. Ashok regretted not buying a better one and realized that the actual problem he would face was that all three uncles had sons overseas, all IT professionals, which would make it almost impossible for him to sustain the competition were he to remain in India. He thus decided to migrate to Australia.

Chapter 3
Selling "Bodies" and Selling Jobs

I first ran into Uday at Advance Technology Institute, a small IT consultancy cum training institute cum software outfit—a typical Indian body shop in Australia—located above a Chinese-run electronics shop and next to an Indian grocery, in the western Sydney suburb of Ashfield. Dark, well built, and wearing a baseball cap, Uday told me that "I have experienced so many things in life," which I soon recognized as his signature line when he became my key informant and I his loyal flatmate. Uday believed that he had good reasons to claim so, and transnational mobility had been *the* theme of his life in the last years. After getting his master's degree in communication technologies in 1997, he applied for more than twenty jobs in India, but got nowhere. Furthermore, while he was jobless and thus unable to secure any suitable marriage match, his younger cousin Ashwin got married, immediately after getting his H-1B visa. Uday belonged to the Patnaik caste where it was a serious indiscretion for younger paternal cousins to marry before their elder cousins. Uday refused to attend the wedding. Finally, deciding to try his luck abroad, he asked his uncle for the contact details of Ashwin's brother, Kishore, also an IT worker and at the time in Singapore. The uncle refused. With this standoff, Uday swore to his father that he would rather starve to death than visit his cousin, and proceeded to Singapore on a tourist visa. Luckily he landed a job, and this brought him a new fate. After moving from Singapore to Australia through Advance Technology, he got a call from Kishore—still in Singapore, his contract was up soon, and now he was asking for Uday's help for possible openings in Australia! Over the following two days Kishore called twice; Uday recorded the times and told me with some satisfaction that Kishore had spent at least SGD 25 for the calls.

Uday's higher status, along with his migration to Australia, had boosted his father's confidence that he could secure him a good marriage. At the final negotiations with one potential bride's family, Uday's father said that the girl would have to pay her own visa application fee and airfare to Sydney after the marriage. The girl's father agreed. But when Uday's father mentioned that the airfare was AUD 1,200, and clarified that there was no off-season rate, the girl's father immediately recited a long list of airfares from India to various cities in the United States and, in conclusion, said that AUD 1,200 was unreasonably high for Sydney. The negotiations collapsed. Not the least disheartened, Uday's father leaped into three other negotiations. One girl turned Uday down for a U.S.-based IT groom. Another was lost after her family proposed an Internet-aided video conference between Uday and the girl, but Uday was reluctant to fork out AUD 200 for the equipment. Finally, in May 2001, Uday's father settled a deal. A few days later, Uday received an e-mail from Ashwin, in Boston, that read: "From unknown source I was told you are going to marry. I couldn't wait to tell you my happiness." Uday insisted that I write to this cousin, to boast that *he* was now host to a *Ph.D. student "from England."*[1] He also urged me to be at his wedding: "If you come, all [my cousins] will be shocked—I also have international friends! Now my status is different."

Indeed, the ability to move around the world formed an important basis for IT people's claim for social status. Ganga, a Telugu doctor in Saudi Arabia whom I got in touch with in an Internet chatroom, asked me to help him find a bride in Australia so that he could migrate there. Meanwhile, Ken, a retired school teacher in Sydney originally from Tamil Nadu, was busy searching for a groom for his daughter. When I told him about Ganga, his rejection was outright: "No doctors. We are looking for computer people only." While an IT groom was almost guaranteed a well-paid job and permanent residency in Australia at that time, doctors with foreign qualifications had to pass requalifying examinations to be licensed to practice. I did not relay the gist of this rejection, but, presciently, several weeks later, Ganga informed me that he was thinking of taking an IT course in order "to get out of the present situation and move on." [2] In India, transnational mobility was even more important for IT people's status. During the dot.com boom of the late 1990s and early 2000, many employers in India had to promise to send potential employees overseas within a year or so, before having them agree to sign the customary company bond for a minimum of two or three years. Even that did not stop parents' turning up at the workplace to ask *when* their sons would be sent abroad. Ravi Kolluru, an IT professional in his early thirties, was in 1999 offered the position of senior software engineer with an annual pay of

INR 3 lakhs in Satyam, a leading Indian IT company based in Hyderabad. Not satisfied with this, he left for Australia in mid-2000. Six months
later, he again contacted Satyam, from Sydney, and despite the looming
slowdown was immediately offered a position with an annual salary of
INR 4 lakhs. Ravi Kolluru reckoned that had he been in the United States
he would have got an INR 5–6 lakh package. Thus, India was producing
not only IT labor force, but also the desire to go to the West.

The high rate of unemployment among IT people—which was an endemic, although seldom highlighted, element of the body-shopping business, as well as of the IT sector in Andhra Pradesh as a whole—was another
factor pushing their emigration. The enormous local resources channeled
into the mass production of IT people, including the almost frantic expansion of private technical institutes in the years leading up to 2000, was
starkly manifest in what looked like gross overproduction in the early
months of 2001. My field data, based on the estimates by nearly twenty
IT training instructors, graduate students, journalists, and IT workers,
suggested that as many as 10,000 IT professionals in Hyderabad had been
unemployed for two months or more by August 2001. Out of the fourteen MCA graduates of Madurai Kamaraj University in Hyderabad in
October 2000, for example, only one had found a job by July 2001. And
of the forty MCA graduates of a private college affiliated with the prestigious Madras University in 2001, only eight managed to find jobs within
two months, seven of them as teachers in private intermediate colleges—
producing future IT graduates. Overall, it seemed that perhaps only 40
percent of all IT graduates in Hyderabad managed to find jobs within six
months of their graduation in autumn 2000. Official estimates, though more
optimistic, were by no means encouraging: according to the State Council of Higher Education (2001e), around half of the IT graduates in 2000–
2001 were unemployed, while a senior official of the Commissionerate of
Technical Education put the figure at 30–40 percent.

Unemployed IT graduates moved from one city to another to search for
jobs, typically finding solidarity in small groups of up to five classmates.
In Andhra Pradesh, the beaten track took them from Hyderabad to Chennai, Bangalore, Delhi, and Mumbai. Moving around also provided an escape from the social tensions brought about by unemployment. Said
Ravinder, a twenty-eight-year-old Telegu IT worker: "In the first week [after
graduation], the neighbors ask you why you are home; the second week
they ask how long you can stay home; the third week they ask why you
still live in this world wasting your father's food!" Widespread IT re-
trenchments in India as a result of the worldwide slowdown in 2001 had
created a new category of unemployed IT people. Often, turning up for
work one morning, IT workers were stopped at the gate by the company's
security guards and given their notice of layoff. Unemployed workers set

up mutual-support e-groups to share information on job opportunities as well as their frustrations, while not a few business scams exploited their desperation for jobs.[3] Many who decided to switch from other professions to IT were hit particularly badly as they had quit jobs to study full-time in order to get IT jobs as soon as possible; nor could they simply return to their previous jobs, due to the difficult employment situation in general.

I met some IT hopefuls who had been unemployed for years. Rama Chandria had run a mess (small restaurant) in Guntur after obtaining his master's degree in electrical engineering. In 1999 he gave it up and took a course in mainframe technology, completing it just as the Y2K boom subsided. He then went to Chennai to "try loopholes," as he put it, talking to body-shop operators and training institutes, and hanging around the U.S. Consulate wishing for some miracle. In May 2001, Rama came to Hyderabad to study Java programming, a skill then in high demand, and tried to go to the United States through a body shop. But the Java job market died soon, and so did his plan. Fortunately he recovered most of the money he had paid the body shop. Over the years, Rama continued to live off his parents, occasionally borrowing money from his friends in the United States. When I met Rama in Hyderabad in 2000, he was sharing a two-bedroom flat with Ravinder and four other flatmates—a private college lecturer waiting for a share of his brother's dowry so that he could go abroad, a civil contractor studying IT, and two others waiting for their visas to the United Kingdom and Singapore. Every night, however, there were usually ten men sleeping on the floor, the other regulars including two who had lost IT jobs recently and one doing a "part-time software job" in a local company without pay. Playing cards and watching TV formed the main routine for the day, and two scooters were shared to roam the town.

Many unemployed IT people that I met believed that finding an IT job in the West would be far easier than doing so in India, if only they could afford the journey. This explains why the global IT slowdown did not put a damper on the queues to emigrate. Despite the massive layoffs in the United States, for instance, the INS received 342,035 applications for H-1B visas between October 1, 2000, and September 30, 2001—a 14.4 percent *increase* from the previous fiscal year.[4] A Hyderabad-based body shop generally took from two to eight months to obtain an H-1B visa, and the waiting period became even longer and outcomes less predictable immediately after September 11 as some U.S. firms ordered criminal background checks on prospective (and current) employees from certain countries, particularly if their work involved Internet security and network-system support.[5] Thus, in the slowdown, a large group of unemployed IT people in Hyderabad who were waiting to go abroad were turned into long-term full-time potential emigrants. Another side to the unemployment-driven emigration was the so-called reverse exodus of IT people to India during the 2001 slowdown.[6] This trend was probably much smaller than reported by the media. Compared to the unhappy ranks of locally retrenched IT people, those who lost jobs and were retuning to India during bad times were roundly taunted: "If you can't find a job in the United States, how can you find one in India?" Besides, "What would the neighbors say?" was an only too real question for IT returnees and their families. A *mandal* (subdistrict) revenue collector in Nellore district, southern Andhra Pradesh, and a distant relative of Vijay Naidu's, had somehow collected INR 3 lakhs to send her husband to the United States in 2000. So optimistic about the future, the officer quit her job, only to find her husband out of a job in 2001 and wanting to come back. She reportedly threatened to commit suicide if he did so.

Fee-Paying Workers and Body Shops in Hyderabad

While the craze for going abroad brought about the flourishing body-shopping business in Hyderabad in the boom years, full-time (unemployed) potential emigrants became its sustaining saviors during the slowdown. Body-shop operators in Hyderabad made part of their profits out of the commissions paid by their overseas counterparts: in 2000 they charged a lump-sum flat fee for each worker placed in the United States that ranged from USD 500 to 5,000, and in Australia, from AUD 800 to 3,000. The other type of commission, per-hour pay, was calculated as a percentage of the worker's salary and payable for the number of hours worked for the first one or two months; in 2000 and early 2001 the commission rate ranged from USD 2 to USD 10 per worker-hour. For India-based body shops, per-hour commissions were more lucrative but also more risky: a worker overseas might run away or the overseas counterpart might just stop payments midway. Body shops in Hyderabad therefore came up with various counterstrategies such as registering joint ventures with the overseas agent and sending workers out as employees of this joint company. By doing so, the India-based body shop could monitor the worker's salary since this would be credited to the joint venture rather than the overseas agent as well as cultivate close relations with workers, who could clue in the body shops on the situation overseas and also generate additional business.

Ultimately, however, it was not commissions but the fees charged to workers that provided the largest and most secure source of profits. In 2000, body shops charged workers INR 1–3 lakhs for being sent to the United States, INR 1–2 lakhs for the United Kingdom, Germany, and Australia, and INR 80,000–INR 1.2 lakhs for Singapore, the range in each case also taking into account various "supplementary services," such as

43

providing reference letters at INR 15–20,000 each, or fake certificates. Workers without a strong IT background paid the most to purchase all the necessary documents.[7] While body shops in India had to come up with special tactics to ensure their commissions from agents overseas, they were in full control of securing payment from workers. If workers failed to meet a payment deadline, they could simply cancel their visa applications. Body shops in Hyderabad thus made their earnings not so much by selling bodies (workers) to clients overseas, but by selling jobs or life opportunities abroad to workers.

During the slowdown, body shops charged workers more, rather than less, partly to set off their drop in income due to the decreasing numbers of workers that they were able to place: "It is harder to get a visa now, so you have to pay more!" a long-term potential emigrant explained. Two final-year BCA students told me how a body shop had wanted to charge INR 6 lakhs each for sending them to the United States, and that even after one called in his father to negotiate, had only reduced the fee to INR 3.5 lakhs. "This is the fate of being an Indian," was the reply when I asked why they considered paying so much. One excuse for charging workers more was the increase in the H-1B visa application fee from USD 600 to USD 1,100 in late 2000. Not sure whether they could place workers quickly and start getting returns, body-shop operators in the United States wanted to shrug off all costs; and when they began demanding an "INS fee" from agents in Hyderabad for each H1-B worker they sponsored, this was, of course, passed straight on to the worker. Some body shops in Hyderabad added their own mark-up and I was told that some even attempted to charge an INS fee for Australian visas!

The overproduction of IT labor also made it possible for body shops to make fat profits by charging workers even for jobs in India-based companies, particularly during the market slowdown. Not that there were more openings in the local market; quite the contrary, placements had become so scarce that unemployed workers were willing to pay high prices to, literally, buy local positions. Almost all my IT graduate informants were aware of such practices and reported fairly consistent prices: INR 1.5–2 lakhs for a job in a company with one hundred to two hundred employees in mid-2001, of which body shops usually took INR 30,000–40,000 and the company's human-resource-department personnel the rest. Workers with more experience and connections approached companies directly and if selected, might have had to pay those in charge of recruitment around INR 60,000–70,000.

Apart from selling jobs both overseas and in India, some body shops— to survive the slowdown—created "training" and "software development" jobs within their own consultancies and sold them to unemployed IT people. A six-month training attachment in Hyderabad—the use of the term "training" merely justified the charging of "training fees" and rarely involved any teaching or learning—went for INR 1 lakh, and the worker received a monthly stipend of INR 2,500–3,000, which means the net price for the job was INR 85,000–92,000 plus interest.[8] After the six months, a worker either was absorbed by the body shop as a legitimate employee, or had to go home. Although some were refunded when they had to leave, the refunding was never more than half of the fees they had paid. These job attachments were advertised in the media as "work on self-support basis with chance to earn," or "career-oriented training," or "on job training"—sometimes more truthfully mistyped as "no job training."

M-Station Ltd., a body shop in Hyderabad run by Sai, a Telugu man in his forties, illustrates well the benefits of creating and selling in-house jobs. Unable to locate any overseas job opportunities after October 2000, M-Station was in crisis in early 2001. In February, Sai created and sold three four-month attachments for INR 70,000 each. Since the U.S. economy was anticipated to recover sometime around July 2001 at that time, those buying the positions were hoping to be sent there soon. Two more positions at M-Station were sold in April, and another three in June. For Sai, these workers' fees alone brought in a sufficiently large amount of cash to stave off closure, and, more critically, they actually enabled him to venture into the business of software-development. M-Station soon secured two software development projects from the state government and one from a local bank, because Sai, notwithstanding his connections as part of the Kamma community, was able to charge very low prices and not ask for any payments until the projects were completed. With the cash flow and skills generated by fee-paying workers, M-Station later even experimented with software-package development—a breakthrough business undertaking for any IT company, requiring substantial investments, particularly in marketing. But Sai was not worried; as he blatantly put it, the skilled labor he depended on would be paying for itself: "If we can sell it, it's the best; if we can't, it is fine. We have nothing to lose. . . . When I train them [the workers], when I place them, when I give them the references, I can always charge them." For the same reason, the resource from free or fee-paying workers greatly facilitated the entry of small IT firms into the international market. InfoGlobal, a medium-sized company in Hyderabad, created and sold about twenty positions in 2001, promising to send all its workers to the United States one day. With these takings in hand, the manager financed a trip to Australia and secured a deal with a Melbourne-based company. Soon afterward InfoGlobal moved from the Ameerpet area of Hyderabad, where most small IT firms were concentrated, to Madhapur, the posh, high-tech suburb of big players.

Finding it a little hard to appreciate fully why workers chose to buy jobs at such high prices, I raised the question with a senior government official, a special advisor to the chief minister for the IT industry, who replied, "What is the alternative?" One fee-paying informant, certain that his career could never take off without prior work experience, told me that this was his "only way to make a break." Vijay Naidu bought a position in a company, Mega Capital Ltd., that his brother-in-law had connections with: "It's not too bad. As long as you don't put yourself in debt, it's not too bad to join the company. At least you have a job. . . . If *you* were a gold medalist in college and didn't have a job for two years, you will know [how it feels]." (Table 1 demonstrates how Vijay Naidu managed financially.) Seeing that body-shop operators were able to make quick money and keep afloat in tough times, some unemployed IT workers decided to start body shops after failing in their attempts to sell themselves or buy into jobs. When I met Rama for the first time, he had given up on learning any new software skills or applying for jobs; instead he had decided to become an IT entrepreneur and had started an unregistered body-shopping business together with Ravinder in working toward this goal. Rejeshekhar, whom I met in Kuala Lumpur, went to Malaysia with his brother-in-law through a body shop run by his friend in Hyderabad. The friend, now a rich petty businessman, had himself been a long-term unemployed IT worker and a victim of an unscrupulous body shop. Sent to Malaysia in 1999 on a visa that did not allow him to seek employment, he only realized his situation after arriving in Kuala Lumpur and had to survive on odd jobs after turning to the Indian community there. A few months later, he returned to India and set up a body shop in association with a Tamil-run labor agent that he met in Kuala Lumpur; within seven months, he sent seven workers, including Rejeshekhar, to Malaysia.

Unemployed IT professionals also acted as subagents for body-shop operators by helping to recruit workers, a role that had its pitfalls. In 2000, Ravinder could no longer endure the social stress of being jobless and borrowed money to go to Australia on a 456 (business) visa sponsored by Puli Reddy, the operator of the body shop Puli Reddy Consultancy in Sydney. After being benched for more than six months, Ravinder saw little hope of recovery in the Australian IT labor market and decided to return to Hyderabad in early 2001 to recruit workers as Puli Reddy's subagent. The job appeared simple: look for workers who were willing to pay a fee to go to Australia, and fill out their visa application forms with embroidered details on their backgrounds, specializations, prospective positions in Australia and so on—it was up to Ravinder how to make their applications look good. Apart from the promise of commissions from Puli Reddy, Ravinder had his own agenda in setting up as subagent, which was to begin his own body-shop operation later with Rama Chandria, using Puli Reddy's contacts. When Gopal, a schoolmate of Uday's, approached Ravinder for an Australian visa sponsorship, Ravinder charged

Table 1
Vijay Naidu's Financial Balance Sheet 1997–2001 (in INR)

Year	Saving as of end of previous year	Income	Expenditure
1997	15,000	108,000 (salary as civil engineer in large state-owned company, and profits contracting a small construction projects from the district government)	60,000 (subsistence; equipment for the construction projects)
1998	0 (all submitted to the parents)	48,000 (salary)	55,000 (subsistence)
		60,000 (cash part of dowry)	30,000 (IT course fees)
		50,000 (given by wife's family)	
1999	73,000	20,000 (borrowed from brothers)	36,000 (subsistence)
			12,000 (furniture-trading business investment)
			35,000 (IT course fees)
2000	10,000	24,000 (from wife's family)	60,000 (relocation and subsistence, own and brother's)
		50,000 (bank loans to do civil project)	100,000 (buying the Mega Capital job)
		40,000 (from parents)	20,000 (equipment and bribes for bank loan for contract projects)
		20,000 (field loans)	
		25,000 (brother's income)	
2001	0	15,000 (stipends from Mega Capital)	40,000 (subsistence)
		35,000 (brother's income)	10,000 (bribes to government officials for contract projects)
		20,000 (profits from contract projects)	
July 2001	20,000		

him INR 1.5 lakhs and showed him all the other forms he was filling out as proof of his wide connections with Australian companies. In June 2001, Puli Reddy asked Ravinder to collect the fees of INR 80,000–100,000 from each worker on his behalf. Ravinder saw this as a sign of Puli Reddy's increased trust. On Christmas Eve 2001, Ravinder went to Australia, still under Puli Reddy's sponsorship, hoping to renew his visa and become a formal business partner of Puli Reddy's. At the airport in Sydney, he was detained. It transpired that Puli Reddy's consultancy had been on the blacklist for six months, and all the visas it sponsored, including Ravinder's, were canceled. Finally understanding that this was why Puli Reddy had asked him to collect the workers' fees in India, Ravinder broke down. After calling everyone that he could think of in Sydney to no avail, Ravinder was deported. I was back in Oxford by the time I heard the news and tried to contact him by phone but failed. Later I was told that Ravinder was in hiding because all the IT guys who had paid him the fees were now chasing him looking for a refund!

India as the Nexus of Global Body Shopping

The large number of IT professionals produced in Andhra Pradesh had turned Hyderabad into a capital node in global body-shopping networks. At the peak of 2000, Aditya Enclave, a three-block, six-storied residential complex on Ameerpet Road, accommodated perhaps one hundred small IT consultancies, whose business invariably rested mostly on shopping IT people. Moreover, based on the numerous body shops and their dense connections with IT workers, training institutes, and other related agencies, Hyderabad had become a geographic coordinating center in the global circuits of body shopping, where information about the global market was ceaselessly exchanged, networks constantly extended, and IT people's career strategies and migration decisions formed.

The coordinating reach can be demonstrated by how Sai's M-Station in Hyderabad put through a deal for forty workers on a nine-month-long project in the United States. In mid-2001, a body-shop operator in Hyderabad received information from a relative working in the United States that a company urgently needed forty workers with ERP (enterprise-resources-planning) skills. Although the requisite workers could easily be found in India, the H-1B visa process would take too long, so the body-shop operator called upon Sai for help in recruiting Indian workers already in the United States. Working through M-Station's associates and former workers in the United States, Sai found a group of forty recently laid off H-1B workers and arranged for one of his associates to be their new sponsor. In the meantime, Sai paid the body-shop operator in Hyderabad a commission of USD 5 for every hour put in by each worker in return for his staying out of the deal. In the United States, the workers' wages were paid to M-Station's associate, who in turn sent them on to Sai. Sai decided how the profit would be divided. Sai brought this U.S. associate into the picture because a nine-month-long project would necessitate regular and face-to-face communication between the client and a representative of the workers. The particular associate chosen for this role had hardly any connections with the workers or the actual employer in the United States and therefore was unlikely to edge M-Station out of the deal. Seeing my amazement at how he monitored such a deal sitting in a dim room located in a somewhat shabby corner of Aditya Enclave, Sai said:

> Forget where I am! If I sleep in the same house with my partner, he still can cheat me. . . . Of course he [the U.S. associate] can get around me. But he has to think about the future. Price is a factor. For [placing] one worker, if I get USD 1,000–2,000, I will be happy; this is very low [a cut] for him. But the most important thing is the time factor. Without me, no way can he get the right people quickly.

Ever alert to the remotest leads for business openings, Sai quickly took the opportunity to sketch out the idea that I could join M-Station as its "special consultant for the Asia Pacific market" and break into the South Korean market (thinking that Korea and China used the same language). As a start, he would cut deals with the clients I introduced and reward me through commissions; later, I could operate my own body shop and decide when and whether to pass on a deal to him, and on my terms for commissions. Sai thought the collaboration would certainly work well: an Oxford-based agent would attract the attention of Korean clients, while I would be completely dependent on him to recruit workers. The overproduction of a flexible IT labor force was what made body-shop operators like Sai so confident about lighting on any opportunity to expand business.

Compared to their counterparts overseas, body-shop operators in India had deeper relations with the workers they handled—always an important channel for future connections for expansion. As one operator defined the difference, overseas body shops were "functional" whereas India-based ones had "emotional elements." While waiting for their visa applications to be processed, usually over months, anxious workers visited body shops constantly. In Hyderabad, long-term unemployed IT workers often gathered in the body shops in Ameerpet to kill the hours in the day, chatting and exchanging information. Syed, a Muslim who operated the body shop Zentech, kept up regular contact with more than twenty workers whom he sent overseas, while some laid off abroad turned to him again for help. Workers also introduced their new friends overseas who wanted to change jobs or migrate to another country, and Syed collected two hundred résumés in this way over a year and a half. He had started sending workers to the United Kingdom in 2000 precisely because the friend of a worker he had sent to the United States wanted to move there.

Connections with specialist IT firms in India also provided valuable links in the expansion of body-shopping business, particularly to countries where there had been few preexisting contacts. Syed, for example, was exploring the opportunities for body shopping in Peru through a friend who had been assigned there by Aptech, the second largest IT training company in India. The Netherlands, too, was still an unknown market for body shopping, and Syed planned to start up a joint venture with a friend who was working for Baan, the Dutch software company, in Hyderabad and had already made regular visits to the headquarters in Amsterdam, to supply workers for Dutch clients. Some IT specialist companies in India collaborated with body shops as a built-in business, sending out selected staff to meet forward orders from India-based or overseas body shops. Most workers would remain in the employ of the company, but their salary, plus an overseas stipend, was paid by the overseas body shop. This recruitment arrangement was cost effective for body-shop operators as these sums were generally lower than the average salary of those whom body shops recruited individually, and more importantly, the company staff were usually better qualified than those recruited through open advertisements. The drawback was that these workers had to be paid their salaries even while being benched; for this reason body shops in Sydney had largely abandoned this collaboration by 2001 as a response to the slowdown.

The large number of reputed IT training institutes in India was also instrumental in reinforcing India's position as a coordinating center for global body shopping. It was a common practice for IT workers in Australia, particularly those with relatively secured status such as permanent residency or multiple-entry visas, to return to India for further training before moving on to a new country. Not only were courses in India ranked far higher in quality and regarded as more internationally competitive, as they were based on the U.S. curricula, but also, a course that cost AUD 300 in Hyderabad cost more than AUD 2,000 in Sydney. I-Logic in Hyderabad had four hundred students in mid-2001, including more than fifty returning overseas workers, over thirty from the United States. There were even a few students from Africa, mainly Kenya and Sudan, who came to know of I-Logic through friends who had studied in Osmania University on Indian government scholarships. When in India, these workers naturally contacted body shops for information on overseas job placements.

Training institutes were sometimes involved in the placement business, functioning much in the same way as body-shop operators. Global Intelligence, an Indian-run IT training company registered in the United States but operating mainly in India, was associated with various body shops overseas, including Advance Technology Institute in Sydney. Global Intelligence in Hyderabad charged particularly high tuition fees because it undertook to place students overseas on completion of their courses. In 1999 twelve students paid INR 200,000 each for an SAP course:[9] three were placed in Australia after paying an extra AUD 2,200 for visa applications and airfare; eight went to the United States. The success of their training cum placement business in India had prompted Global Intelligence and Advance Technology to set up a joint venture, AGI Knowledge, to develop the business on a larger scale, though this did not work out due to interpersonal conflicts. Glogo Consultancy, a medium-sized company in Sydney engaged in both software development and body shopping, mainly recruited workers through an IT training institute in New Delhi formerly run by Glogo's founder, Samy, and now, by his brother-in-law.

The fact that India was also a regional center for IT training catalyzed connections between IT people from different parts of South Asia, and among IT professionals of Indian origin from different parts of the world, which again enhanced India's superior position as the global coordinating center for mobile IT labor. For example, Nambi, a young Tamil from Colombo, Sri Lanka, moved to Chennai in 1994 to do an MCA course and an IBM training program. According to him, his greatest accomplishment in doing so was the "IT friends" that he made; consequently, he had about ten close Indian friends spread across the United States and Europe and in Singapore. In fact, it was a close friend in Singapore who put Nambi in touch with the body shop in Australia that arranged his move from Sri Lanka to Sydney. A substantial number of Bangladeshis also pursued higher education in India including some who went overseas through body shops after graduation.

Apart from the dense connections within the IT community, a large number of conventional emigration agents in India also expanded their business into IT placement, thus bringing to the body-shopping circuits their well-established networks and migration expertise. Rama Consultancy, an agent specializing in permanent migration run by a Telugu based in Hydrebad, started IT recruitment in November 2000, using its existing infrastructure. Its four full-time employees in Hyderabad handled recruitment while its six employees in Chennai, where most foreign consulates in southern India were located, processed migration paperwork. At the time of my interview, Rama Consultancy had placed two IT workers from the Middle East in the United States and Malaysia, and one from Singapore in Australia—directly, without having to return to India first—because

51

Rama coordinated everything through his connections in Singapore and Australia. Rambabu in Tanuku town was an unusually shrewd business-man. When he was seventeen he converted to Christianity out of an "in-terest in Western culture," and in 1995 went to London on a missionary exchange program to study theology. In the time away from Bible study, Rambabu visited immigration solicitors in London extensively—to find ways to stay on in the United Kingdom. After four years, Rambabu re-turned to India, resigned from the priesthood, and became a migration agent. Initially he took on permanent migration cases, then in 2000 shifted to IT recruitment. During the slowdown, he continued body shopping to the United Kingdom but also started to send female domestic workers to the Middle East. Conventional migration agents also ventured into IT recruit-ment by collaborating with body-shop operators. For example, an estab-lished Delhi-based consultancy run by a Sikh specializing in emigration to Canada approached Syed at Zentech in Hyderabad for workers in 1999, soon after developing connections with IT firms in Canada.

India's role as the coordinating center of global body shopping reflects the simple fact that the profits and advantages accruing through the pro-duction and supplying of Indian IT labor to the global market were large enough to be shared by varied actors (e.g., body shops, specialist IT com-panies, and training institutes). Body shops and the related players consti-tuted an "informal IT sector" where economic transactions were hardly reflected in official records and involved various unlawful activities such as making false declarations and issuing fake documents. This informal IT sector was, however, an integral part of the formal IT sector both in India and abroad: some IT workers obtained their first work experience at body shops to make them a "ready" workforce for the industry, and more importantly the majority of IT professionals in Andhra Pradesh en-tered the international market through this informal sector. Unemployed workers provided cheap and flexible labor for the global market, but more critically they formed the resource base that allowed body shops to survive the market downturn and move into the formal sector by grow-ing into viable software firms.

Chapter 4
Business of "Branded Labor" in Sydney

At the time of my fieldwork in 2001, most body-shop operators in Sydney were those who migrated to Australia as permanent settlers between the 1980s and the early 1990s[1] and had switched into the IT sector from other professions (particularly engineering). They had intended to set up businesses in software development and were pushed into body shopping after facing difficulties in expanding the Australian market for software products. Body shops in Sydney thus clearly differed from their counterparts in Hyderabad in the trajectory of development, but they also shared crucial commonalities, particularly the overlapping between body shopping and other business operations. This chapter traces how the body-shopping business emerged and evolved in the Indian community of Sydney.

Most body-shop operators in Sydney had gone into their own businesses because they felt that their career prospects were being frustrated by "glass doors" that blocked their entry into managerial ranks from technical positions, and glass ceilings that prevented their promotion beyond a certain level (Simmons and Plaza 1999; Watson 1996; Saxenian 2001). The overwhelming majority of permanent migrants with IT careers that I interviewed—basically the same group as the body-shop operators—were in middle-level positions such as project leader, senior programmer, and chief programmer/architect/analyst.[2] Karthik, a Tamil IT professional working for Ozmails, a large Australian communications company, complained that being stuck in a position as a middle-level technician, Indian IT people had "no place to go": "Marketing is everything here. If you don't know about marketing, you can't go up." Khrishna from Kerala had worked in Australia for more than ten years and was bitter about his dead-end career: "What are sales and business? They are lies! They have nothing to do with the brain. No substance!" Many informants were upset that their master's or Ph.D. degrees were given no weight at all by their bosses.

Both my Indian informants and their Australian colleagues were keen to point out that upward mobility of many Indians was impeded by a lack in communications savvy. The largely horizontal networking organizational pattern in the IT sector in some sense, therefore, was seen by many as more difficult for Indian professionals to negotiate, such as Stephen Loe, a veteran in the IT recruitment business and a manager of Morgan and Banks Technology, one of the largest IT recruitment agents in Australia:

> In Australia, organizations are very flat. You may run into your boss any time. You may work the next door to the director general. Aussie bosses may ask: "Hi, Biao, how was your weekend? How is your wife?" Indians don't know how to answer these questions interestingly. Then the boss asks: "By the way, I have this problem, could you sort it out for me?" These are opportunities! Indians are not free-flowing, not smooth with things.

Golf courses, pubs, or bars, important venues in which corporate information is routinely exchanged and the "real deals" in office politics made, were not amenable settings for the many vegetarians and teetotalers among the mainly upper-caste Indians who arrived as permanent migrants before the mid-1990s. Regular office gatherings were often a cause for anxiety, as with Sree Kumar, Uday's friend, who asked me for tips on how to survive an upcoming party organized by his company. Even Uday spent his coffee breaks avoiding his colleagues because he could not handle Australian humor; instead he chatted on the Internet with friends in India and elsewhere, and briefed me daily on what was happening with his friends in New York or Singapore. Quite a few Indian and Australian informants suggested that Indian IT people did well in software programming precisely because this area required little interpersonal communication.

What Indian IT professionals were perceived to lack, of course, was merely the communication skills used in the (culturally) Anglo-Saxon-dominated corporate world: the interactions I witnessed at public activities organized within the Indian community—"ethnic functions" as they were referred to by my Indian informants and the Australian media—were often far more lively than in most Australian pubs. The perceived lack of communication skills thus may just as well be the lack of willingness and interest on the part of managerial staff to reach outside their ranks. No matter what the reality is, it is clear that "ethnicity" remains an issue, and covert discrimination exists in high-tech workplaces. Nandan Desai, a manager with Mastech Australia, the Australian subsidiary of the U.S.-

based IT service firm, told me that they had to recruit "white" candidates from the United States and United Kingdom for senior-manager posts because workers in Australia were uncomfortable with "brown" bosses.

While various career and workplace frustrations motivated many to pull up stakes and move elsewhere (mainly the United States), for those who chose to remain in Australia, the sudden and global layoffs during the recession of the 1990s proved the decisive push to becoming self-employed. Chandary, now a well-established Telugu businessman, was one such professional: "I could look for another job. But the recession may come back in another five or ten years. We are outsiders. When the business gets bad, we are always the first to go." He thus decided to set up CSR Holding Pty., which initially specialized in programs for resource management in automobile plants but later focused on body shopping. The experience of the recession had perhaps made the Indian professionals more psychologically and philosophically prepared to accept risks. Like Sitaram Reddy, who migrated from Andhra Pradesh to Australia in the late 1980s, and who started up a business in 1993 with his wife after getting laid off: "You have to take a risk whatever you do. I and my wife were working in Australian companies. When there was not enough work to do, we had to go as well. The same thing!"

Most of these would-be technopreneurs started in software services—designing, implementing, and maintaining programs for clients, somewhat similar to contract consultants—and adopted various strategies to ensure business security such as starting small to reduce risk, and moving back and forth between self-employment and full-time employment. Another common precaution to ensure the financial security of the family before going into business was investing in real estate. Investing in houses made sense also because owners could apply for home loans, which could be as high as 90 percent of the value of the house. Home loans were particularly important, as most Indian IT entrepreneurs seemed very reluctant to borrow money from other channels. The procedure required by banks for commercial loans was seen troublesome and paying interest as not worthwhile since their fledgling businesses were often ad hoc operations. At the same time there was a deeply ingrained cultural stigma about borrowing money from fellow Indians, which was compounded by the fear of failure in business: How would one face the creditor in the community if one failed? Business plans, therefore, were tailored to fit immediate financial capacity to minimize any such social repercussions.

It would be impossible to overlook the significance of wives' incomes as an important source of security. All the thirteen body-shopping businesses that I investigated in Sydney, except for one, were set up by married men, with the majority of wives working in the public sector (six in government departments, three in schools, one in a private company, and the other two self-employed assisting their husbands). This was no coincidence; public-sector jobs paid less but were far more secure than private-sector ones, as Chaya explained: "The family needs some money coming in on a regular basis. Otherwise it would be too scary for us." Piranavan, a Tamil IT professional in his forties and the owner of Osin System, a consultancy focused on programs for large data-set management, also stressed that the regular work hours meant that his wife could ensure the routine of a stable family life while he was on the go. Khrishna had his wife switch her job in a small private company for one in the state taxation office when he began toying with the idea of setting up a business—though he never did. Ample literature has highlighted women's often invisible contributions to immigrants' businesses with their wages from paid labor in the mainstream economy, say, as factory workers (Bhachu 1988; Werbner 1988); in this aspect the owner-operated Indian IT business were not so different from the street-corner groceries.

Some of the one-person companies did expand, by hiring, in most cases, Indian IT workers. Co-ethnic workers were an important resource, not for any supposed "attachment" (Waldinger et al., 1990, 37), but because they permitted special flexible employment relations. Sitaram Reddy, the owner of Softworld Consultancy and the only IT entrepreneur that I met who was adamant about not venturing into body shopping, had taken on four workers at the time of my fieldwork. Until 1995 Softworld was a one-man outfit, but when Sitaram Reddy secured a large development project from Compaq, he roped in his wife, and enlisted a friend for AUD 50 per hour. After completing that project, he secured an even bigger one from Compaq, for which he hired three fellow Indians in Australia, one full-time and two part-time. One of the part-timers, Kondepudi, had quit his job in a large company to take on freelance consultant work for Sitaram Reddy and other clients. The relationship between Sitaram Reddy and these workers ceased after late 1997 when many IT projects were put on hold due to the Y2K problem. With the market recovery in late 1999, two of the former employees returned, and, predicting a strong market for the year 2000, Sitaram Reddy recruited two more workers, also Indians. Furthermore, he paid all the workers fixed salaries for the year 2000 instead of project-based remunerations—fixed salaries were more economical if he could get projects continuously throughout the year—but made clear to them that the manner of payment for the following year would depend on the market situation. Flexible employment made some IT professionals employees and employers simultaneously, as when Kondepudi worked for Sitaram Reddy while at the same time employing other Indians for his own IT consultancy projects.

Flexible hiring could also be combined with on-the-job IT training. In 2000 Piranavan engaged the services of two full-time but nominally paid workers and two unpaid part-timers. None had prior IT work experience and saw the hands-on experience they gained as the main reward; for Piranavan, training them took up his time but saved a substantial payroll. Before these workers, Piranavan had trained six Indians since 1997, four of them women. According to him, all had found jobs in proper IT companies after gaining experience with him.

Similar to these flexible employment arrangements, ethnic-network-based subcontracting paid by the piece was another means by which the small IT businesses expanded. Raj Electronic Consultancy, owned by Raj, a civil engineer with a keen interest in IT, provided services related to security systems. When asked whether he employed workers, he answered "yes and no." Because Raj's projects involved a range of skills, he divided them into subprojects and contracted them out to his five associates, four of them Indians, including three software professionals and two electronic engineers. Raj's wife ran a home-based kindergarten, mainly for other Indian families living nearby, and Raj sometimes sought associates through his wife's networks. In another case, Rama, an Indian-Fijian who used to work in the government of Fiji and now ran a spice shop in Sydney, got to know quite a few Indian IT people as customers of his spices. Influenced by them, Rama set out to export computers from Australia to Fiji, mainly to the government departments that he had connections with. Then some IT friends suggested that he could charge more if he loaded the machines with software programs catering to the specific needs of the government departments. This idea worked; and Rama outsourced the task of customizing and installing the software to his Indian friends. By 2001 more than ten Indian subcontractors worked for him.[3]

"Marketing and Development Are Totally Different Stories"

While owner-operated IT businesses could be expanded relatively easily through flexible hiring and subcontracting, growing a market share proved to be far more difficult. Initial market entry was oftentimes gained through having one's former employers as clients. For example, Piranavan, who set up his business in 1991, continued to work part-time at the insurance company where he had been employed for four years, not just for the sake of economic security but as a channel for getting projects. Others started out as subcontractors of their former employers. Leela, a Tamil Brahmin with a large extended family in Sydney and the only Indian female entrepreneur that I came across, used to work in an insurance company as well. When she came to know that the company planned to upgrade its IT system, she quit to set up a consultancy to provide software design and implementation for them. She brought in eight workers through a body shop in India for the task, and when the project was completed, she placed the eight workers in other companies and thus started body shopping. Surprisingly, small firms only occasionally found clients through fellow Indians, and it was commonly held that the large presence of Indians in the IT industry could not be counted on. Remash, a Telugu freelance IT consultant and also the owner of the body shop WinWin Recruiter, explained: "They [Indian friends in big companies] fear. Say, if you introduce me to your boss, and I can't do the job well, what will the boss say? What will the colleagues say? I am not keen to ask other Indians to *give* me opportunities either."

When Piranavan quit his part-time job in 1994 and moved out of town into a big house to concentrate on his software development, he immediately faced the problem of finding clients: "That was a mistake. I cut myself off from the city. Had I worked for the company longer and stayed in the city longer, I would have had lots more connections." At that time Piranavan's main business revenue came from implementing and maintaining a program he had designed for three companies, each contract lasting eight years. According to Piranavan, this kind of long-term servicing deal was not at all profitable, but, for him, developing the program into a package and marketing it more widely as larger software companies do was not an option:

> Marketing and development are totally different stories. The investment is huge to sell packages. You have to give traders very high margins. Microsoft Windows 2000 is sold for AUD 120 in the market. Microsoft sold it at AUD 85 to my Chinese supplier, and he sold it to me at AUD 115. All the sales take place along chains. How can *we* develop the sales chain?

Besides the considerable financial resources required—the outlay for successfully marketing a software program being perhaps four times higher than for developing the program—the IT people I interacted with in Australia seemed to lack interest in marketing also. Sivakrishnaviram, an IT contract consultant and originally from Tamil Nadu, had developed a software package facilitating medical consultation but had sold merely eight copies (of which six were sold by his brother, a doctor!). But he was more satisfied that the software worked well than upset about the low sales. Sitaram Reddy had a similar attitude:

> I don't want to develop my company to something that big. Lots of social factors will come in when you want to do big businesses. We are not good in that. . . . I always want to do something new. I don't want to stick to one thing and sell it to all the world. *We are professionals.* . . . I love computer networking, not social networking.

My informants' social circles appeared highly homogeneous: mostly limited to those Indians working in IT specialist companies; very few consciously reached out to trade circles or set about targeting potential customers. Vikram, an ambitious young Telugu working in Telstra, the largest communications company in Australia, had very detailed plans for his future software business, but when asked about marketing strategies, he looked puzzled:

> I don't know about marketing. . . . Is marketing important? Marketing is not important. *Brand is important.* If we want to buy cosmetics, we buy those from Paris. If we want to buy electronics, we naturally think of German, Japanese goods. They have captured the market. They don't need any *skill* to market; you can't change that. Now Indian IT people have got their brand. *Clinton* went to our place! We can sell ourselves easily!

Vikram's statements struck a familiar chord: faced with the problem in developing a "brand" for the software products they designed, the small and one-man Indian IT companies went into the business of providing "branded labor" instead. When the demand for Y2K programs bit hard in 1998, small IT entrepreneurs crossed naturally into body shopping when asked by companies in Australia for IT workers or approached by firms in India for information on job openings in Australia. Of course, many more took the initiative to jump into the market.

"Only Indians Can Handle Indians"

The successful branding of Indian IT labor was not entirely to do with their inherent attributes such as technical competence as my informants were persuaded to believe, but was also imposed from the outside. With the ever escalating global mobility of IT workers, employers and large placement agents often relied on stereotypes, particularly when having to quickly screen applicants from all over the world. A manager of a large Australian IT placement agent told me that the applicant's nationality was one of the first things noted when processing an application, and he even went so far as to say that applicants conforming with the stereotypes in the trade would have a better chance in getting jobs. For example: Indians were good at general skills, particularly mainframe technology; Filipinos were good in the C language; and Chinese were better in engineer-related programs—a Filipino or Chinese applicant claiming good mainframe skills would attract more critical scrutiny, or simply be dismissed were there Indian applicants.

The attributes of the Indian IT branding in Sydney were not entirely positive. The Indian IT workers were often thought difficult to manage. Michael, a manager of a large Australian IT placement agent, was married to a Chinese woman and thus claimed to know "Asian cultures" well; he told me that the industry had a "resentment" against Indians now because, "They never work quietly. Working for three months, they start asking for promotions. They always look at other places and want to hop away." Another manager, echoing Michael's view, pinned this down to the widely circulated stories of successful IT Indians in the United States that encouraged high and sometimes unrealistic expectations among those in Australia, and led them to be impatient and demanding.

The negative perceptions of Indian IT workers were so profound as to change the way they were recruited. According to Ravi, a Telugu IT professional and one of earliest in Sydney to venture into body shopping, whereas most international job interviews had once been conducted by telephone, from the end of the 1990s, no Australian or American placement agents dared to pick Indians based solely on this. It transpired that some agents, after going through all the immigration procedures and paying the airfare and other expenses, found that the person who turned up at the office was completely different from the one interviewed. While telephone interviews were still the norm for candidates of other nationalities, particular caution was exercised with Indians. These and other perceptions prevalent in the industry dissuaded Australian placement agents from dealing with Indian workers directly, and, instead, they preferred to recruit those managed by local Indian agents—"only Indians can handle Indians," an Australian placement manager advised.

There were some instances where small non-Indian placement agents had attempted to disregard the reluctance to manage Indians directly. A Frenchman running a small labor agent had wanted to collaborate with Senthil, a former 457 visa holder from Andhra Pradesh who had recently become a PR. He asked Senthil to search for Indians who were working in Southeast Asia, reasoning that those with overseas experience would be more employable than those straight from India. Senthil, however, was eager to bring in his brother-in-law and classmates from India, and, as he put it, their partnership evaporated because of "lack of mutual understanding": "How many Indians in Singapore and Bangkok can you get? [The Frenchman] was always worried—'what about if the candidate can't find jobs?' I told him they can wait. He was always worried!"

The myth that "only Indians can handle Indians," however, did emphasize the special advantages that body shops enjoyed in recruiting Indian workers as many came about just by recruiting "friends of friends." Before Kana and Ravi, the co-founders of Advance Technology Institute, went to Singapore to recruit workers for an order from ICON Recruitment in

1999, they sent e-mails of the details of the skills they were looking for to their friends. The friends in turn forwarded the messages to others in their circle of friends. Consequently, within two days of arriving in Singapore, Kana and Ravi had interviewed twenty suitable candidates and, after four days, settled on six. IT people in Australia and other countries often received e-mails from fellow Indians, bearing the subject lines "work opportunities in Australia" or "want to work in the U.S.?" Likewise, Kana always encouraged the workers he sponsored to bring in more IT people when they went to India on visits, and about five workers at Advance Technology were recruited in this way by mid-2001.

Informal Indian networks not only enabled swift recruitment but, more importantly, were a measure for guaranteeing competence and qualifications, the key to success in the recruitment business. In Australia, workers normally had to show satisfactory performance during a trial period of two to three weeks; otherwise the agent was expected to find a suitable replacement. A worker who failed to satisfy repeatedly had to be sent back to India, which not only lost money and cost time, but also jeopardized the body shop's relationship with the bigger agents they were tied in with. Owing to the personalized connections, the Indian workers recruited by body shops were no less qualified than those non-Indians supplied through big agents. Vand, a twenty-three-year-old IT professional working for Advance Technology, was well known among Telugu IT people in Sydney for his prodigious talent. He was brought in by Vijayarka, a onetime business partner at Advance who had returned to India in 1999 to take an IT course and was taught by Vand, then still a third-year undergraduate student. Six months after his arrival, Vand brought in his friend Maruti, who had worked as a freelance IT consultant and tutor when he was still a college student. Maruti was not sure whether to regret coming to Australia given that the IT sector there was technologically less developed than in India, and he was planning to go on to the United States. Body shops in India also recommended to their overseas associates workers who were unusually highly qualified, as this increased the possibility for earning commensurately high commissions, and at the same time helped raise the reputation of the associate body shop.

IT people themselves sometimes preferred being handled by Indian consultancies. Uday's brother Ashok had three years' experience with Baan in Hyderabad and entered Australia as a permanent resident under the skilled immigration program. Despite his eminently marketable background, he had sought jobs through Indian body shops. Uday explained: "There are many hurdles to get a job through Australian [placement] companies. Indian consultancies will be much quicker. They give you the inside stories: what you should say to this agent and what . . . to that one. They tell us the loopholes."

61

Body shops also made special efforts to help their recruits adapt to the Australian job market, particularly by drumming into them the awareness that communication skills were more important than technical knowledge in finding a job. As Piranavan asserted, Indians frequently failed interviews due to a lack of communication skills rather than technical know-how: "No one can tell how good your technology really is in one hour's time [during the interview]. You have to show you are confident, have a positive attitude. You have to explain things clearly to the clients. Communication is the first thing I teach when the guys are here." Some body shops in Sydney offered an orientation course on "Australian work culture." Coaching covered the essential dos and don'ts: how to say "How are you doing, mate" in the Aussie way; how to rid curry smells with certain types of perfume or cologne; how not to condition the hair with coconut oil (routine in southern India); and how to dress for interviews—common recommendations included a yellow tie with a blue shirt for middle-aged interviewees, and a red tie with a gray shirt for younger ones.

Overlapping Businesses

Although "forced" into the business of body shopping, the operators often in turn used the business to facilitate other IT operations. In the years 2000 and 2001, a significant number of body-shop operators expanded the scope of their businesses to include IT training. Body-shop operators were keen about IT training first of all because, just like body shopping, the business did not entail any prior market penetration. The IT training courses in Sydney managed by body shops were highly "ethnicized": courses were taught by Indians, targeted at Indians, and generally advertised through Indian newspapers, posters in Indian groceries, or at stalls set up at ethnic functions—though, of course, instruction was in English. As a result, for example, 80–90 percent of the trainees at all the courses offered by Advance Technology in 2001 were Indians. Most of those who signed up for IT training courses were, similar to the early body-shop operators, those who arrived with professional backgrounds other than IT. Many were unemployed, or more likely underemployed (e.g., mechanical engineers working as truck drivers; or a person could not find a full-time job despite a willingness to do almost anything). Although the underemployment of professionals did not make Indian immigrants any different from many others, Indians seemed to have a heightened sense that IT skills would offer a way out of their predicament. Imitaz, an architect taking a course at Advance Technology, had designed some ten buildings in Mumbai but could not get the necessary approvals to undertake independent design projects in Sydney. An Indian colleague of his who faced the same problem had shifted into an IT career path and, within two years, was making double Imitaz's salary: "I can't say there is no discrimination in IT. But it is much better than other professions. IT people are so badly needed that they *can't afford* to discriminate!" Singh, a civil engineer with a reasonable-paying job, decided to move into IT as well:

> When you are young, you don't care. You do anything you like. But when you have a family, you have lots of responsibilities. I must change my career now. . . . Doctor's career is measured by five or ten years— you are promoted every five or ten years. For ordinary engineers, the pace is two or three years. For IT, things are measured by *months;* anyone can move up quickly.

The ethnicization of IT training courses in Sydney went hand in hand with transnationalization. In most cases, the instructors were sponsored straight from India, which for body shops not only meant they were cheap, but also that they were, almost always, multiskilled. An Indian instructor could teach three or four different programs at the same time, a capability not easy to come by in Australia. Besides, the course content was normally cloned from those in India, which were thought to be of a higher quality. NewSkies, a training institute cum body shop operation run by a couple from Karnataka in Sydney, was in collaboration with WinIT, a well-known training institute in Hyderabad. WinIT provided them with its courses and two teachers for a fee. This affiliation was highlighted in NewSkies advertisements in Sydney, which attracted immediate attention as many Indian immigrants knew not only about WinIT but also about the two trainers through their own connections in Hyderabad. This collaboration continued even after WinIT was closed down in Hyderabad.

All the Indian-run IT training courses emphasized that they were "employment oriented," or as some consultancies claimed "employment guaranteed"—if trainees could not find a job after the course, the body-shopping part of the consultancy would find one for them. Body shops sometimes even suggested students lodge their job applications when they started a training course, claiming competence in the skills they were about to learn. The assumption was that by the time an application was processed, they should have enough knowledge to cope with an interview. Moreover, body shops offered reference letters—in the capacity of IT consultancies with multiple business lines—as proof that the students had acquired work experience in teaching, marketing, or administration, whichever was relevant for the job applied for. (In some cases, body shops had willingly stated that the student was one of the founders of the institute!)

IT training was sometimes also taken as the starting point for developing other IT businesses including body shopping. Siva Prasad Rao, a journalist who migrated to Australia in the 1980s as a refugee, could not find a proper job in his field, and he thus studied IT and in 1999 started a home-based course with one computer and one student.[4] Thanks to his background in journalism, his good communication skills soon made him a popular instructor. As some students approached him to polish their résumés and other documents for job hunting, Siva Prasad Rao gradually came to know placement agents. When I interviewed him in 2001, he was planning to recruit from India candidates with good communication skills and a solid grounding in mathematics, provide them training in the skills needed by placement agents in Australia at that moment, and then place them out upon completion of the course.

There were other synergies in combining the businesses of training, body shopping, and software development. For example, visas for migrants such as technology instructors who could help increase local skills were treated much more expeditiously by the immigration authorities; thus, body shop operators could relatively easily apply to sponsor workers under the guise of instructors, while, indeed, using these same workers as instructors or in software development when they were benched. Furthermore, both body shopping and IT training could facilitate entry into the market for other IT services and products. For example, an Indian woman working in an Australian company was satisfied with her IT training course provided by ComLink, an outfit that handled training, body shopping, and software development at the same time run by Vinnie, the young businessman whose mother learned of the term Y2K in Hyderabad before he did in Sydney. When her Australian employer sought a training company to provide a group course for its six Australian workers, she recommended ComLink. In addition to that deal, Vinnie won a software-program development project from the same company later on. Ravi used to be the representative of a Hyderabad-based software company to Australia and decided to focus on body shopping, but the Hyderabad headquarters required him to concentrate on software development, and this was why he resigned and joined Kana to set up Advance Technology to specialize in "HR" (human resources), the term Ravi used for body shopping. He explained why he gave so much weight to body shopping:

> Software development is a very responsible job. If a company outsources a task to you, they must trust you a lot. . . . How can you be sure about other people [when outsourcing]? The first option is go for big names, like Lucent or Anderson. They are very expensive. The second option is go for relationships. If we put a person in the client's company, we can talk to them regularly, then we can have relationships and they can

trust us. Even for very big Indian companies, placement still makes up for 60 percent of their revenues. Why? [To maintain] relationships! This is the best way for us [Indian software companies] to get into the market.

Same Roof, Different Hats

Overlapping business activities, more often than not, also represented overlapping business entities that spanned multiple countries. In 1996, Aberami, a university researcher of Tamil origin in Melbourne, went to Singapore on a year-long academic exchange program and, while there, met Rangarajan who was working in a large IT company. They both invested AUD 50,000 to set up a joint software-development company, Singdin Technology. Rangarajan was responsible for software development, and Aberami for marketing as well as overall management because of his connections in India, Australia, and the United States. Hence, Singdin could outsource part of its development tasks to Bangalore and Chennai at favorable terms; for the same reason, Singdin was registered in Melbourne though its main operations were in Singapore. The firm's profit was to be divided equally between Aberami and Rangarajan. In 1998 Rangarajan started his own body-shopping business to bring workers from India to Singapore for both Singdin and other clients; this business was registered as a subsidiary of Singdin. When Rangarajan recruited workers for Singdin— by late 2000 about forty—he did not charge Singdin but instead made his profit by subsequently placing the workers out from Singdin to other companies. This collaboration went well, and in 1999, Rangarajan and Aberami launched another joint venture, with an Indian firm in Chennai this time, to specialize in medical-record transcriptions.[5] All three parties invested equal amounts of money, and this joint venture in Chennai had hired sixty workers by late 2000.

I got to know Aberami through Chaya, a Tamil who first went to Australia as a Ph.D. student in the 1980s and came to know Aberami at that time. Chaya set up Sysway IT Consultancy in 1998 to provide software services for monitoring production processes in mining companies, and when he had difficulties in getting business deals in 1999, he contacted Aberami for advice. Aberami suggested Chaya "join" Singdin as its Sydney representative. Chaya ran this Sydney branch of Singdin independently, that is, shouldering all the investments as well as the possible loss. He would take a fixed monthly salary of AUD 1,000 out of the profits, if he could generate a profit higher than that; the rest was divided in a 3:3:4 ratio between Aberami, Rangarajan, and Chaya. Despite initial difficulties, the collaboration with Singdin enabled Chaya to develop both his body shopping and software development business quickly. Using Singdin's profile—a company with branches in Singapore and Sydney—and its revenue reports, he obtained sponsorships for workers easily; by mid-2000 Chaya had brought nine workers from India, five placed out to other companies, and four to develop an online learning project that he thought had great market potential. With the earnings from body shopping alone, Chaya bought thirteen computers and planned to set up a full-fledged in-house development laboratory by early 2001.

Apart from gaining credibility and an impressive corporate profile, being part of Singdin had other tangible benefits for Chaya, for example the deal in late 2000 to customize an e-commerce software package developed by Singdin in Singapore for the Australian environment. According to the terms of the deal, if Chaya succeeded in marketing the customized program, the profits would be divided in the same 3:3:4 ratio between Rangarajan, Aberami, and himself; if he failed, he had no obligations to the Singapore office. In this mix of internal and external transactions, all parties reached a win-win situation. If Chaya and Singdin in Singapore had been two unrelated firms, Chaya would have had to purchase the original software; equally, if indeed an actual branch of a single firm, Chaya might be less motivated to customize a package and explore the Australian market as the profits might all go to company headquarters in the end. The overlap between Chaya and Singdin also benefited Singdin's other businesses. When in June 2000 the manager of the medical transcription center in Chennai visited Sydney to explore possible cooperation with hospitals there, Chaya used his own connections to arrange the meeting and even accompanied the manager on a couple of visits.

But Chaya did not want to limit his business expansion to being part of Singdin only. In May 2000, using the name of his original company, Sysway, he set up a joint venture with Nadarajah, a brother of an old school friend in India. The joint venture, TrueInfo, started as an IT training institute that was later expanded to body shopping. Chaya put in AUD 7,000 to rent an office and buy four computers, Nadarajah designed the curriculum and did most of the teaching, and both agreed that the profits, as well as all further investments, would be shared fifty-fifty.

Monitoring overlapping business transactions was deliberately left ambiguous. When I tried to clarify the arrangements between Chaya and Aberami in detail—for instance, how the exact amount of the initial investment made by Chaya for the Sydney branch of Singdin was determined—Chaya dismissed my questions: "You can't define everything that clearly. Otherwise we can't do anything. We have to earn some money first, then think about other things." And when I asked Aberami how he could check how much money Chaya made in Sydney to ensure receiving his 30 percent profit share, he agreed that it was impossible to monitor all Singdin's transactions in Sydney, but added:

> I am not worried about this. Number one, I have nothing to lose. I didn't take any money out of my pockets. Number two, we both can have more opportunities and networks by tying up with each other. At least this [having an office in Sydney] is good for the company's profile. If I want to do business in a big way after I retire, the connections in Sydney should be useful. . . . There is no hurry to divide the profit. You can't build Rome in months. Let the business grow first.

Why were overlapping business operations not simply merged into a single business entity? Why, for example, did not Aberami, Rangarajan, Chaya, and Nadarajah form a single corporation as directors holding different share parcels? This apparently concerned the dialectic between the "knots" (business entities) and the "net" (network) in overlapping business cooperations: owners stayed in a particular "net" precisely to develop their own "knots." Chaya admitted voluntarily that he used the name Sysway and not Singdin to set up the joint venture with Nadarajah in order to consolidate his own business base and, eventually, to internalize all his businesses that now overlapped with Aberami's under his own roof. Thus, in ambiguous cooperation, different parties can share limited resources efficiently at the same time as individual owners/entities know where their own interests lie, while a merger would reduce not only the scope for resource mobilization but most importantly the creative autonomy of the "knots." It was crucial, thus, for managing the dialectic between a "net" and its "knots," that transactions retain their individual identities under different (overlapped) names.

This was illustrated by Vinnie's manipulation of his interests under different company names. Vinnie registered his first company, Achieve, in 1997 while he was still working at Unisys Sydney, a global IT services and solutions company headquartered in the United States. Achieve remained on the shelf until early 1998 when Vinnie tied it up with an India-based company for a joint body-shopping venture. Soon afterward, Vinnie quit his job and registered a second company, Vision, with a visiting Indian technopreneur[6] to focus on software consultancy targeting the local Australian market. Vinnie's third company, ComLink, was set up in 1999, also in Sydney, with a close friend of his to do IT training, but in early 2000 the friend left for the United States and Vinnie bought up his share. At this point, Vinnie put all the businesses under one name—ComLink. Internally, the three companies were maintained as separate accounts, but when advertising the services and dealing with customers, ComLink was the only name used. It was Vinnie's goal to internalize the businesses of Achieve and Vision into ComLink, to which none of his partners was opposed: none of them wanted sole ownership of the companies, and they believed that, at least for the time being, using a single name did help make ComLink a known brand in an ever competitive market, and that this in turn helped secure business deals for each company.

Ravi was another body-shop operator who knew well the significance of names. Ravi had a relative who was a movie superstar in India. This conferred on Ravi a high visibility among the Indian immigrant community in Australia, particularly those from the south, and as a result Ravi knew almost all the Indian IT entrepreneurs in Sydney and was also well connected with many in Singapore and the United States. He had registered a range of company names in India, Australia, the United States, and Singapore to play around with in collaborating with others for body shopping—he sometimes even borrowed names from other companies. In early 2001, for instance, he urgently needed to bring fifteen workers from India to Australia. Ravi asked a U.S.-based company run by the uncle of one of his friends that had earnings of USD 18 million in 1999 to provide the cover letter to DIMA seeking sponsorship of the fifteen workers—purportedly for a joint project with Ravi. In this case, Ravi knew that using the name of a firm of this size would add weight to his sponsorship application. This was how he summarized his actions: "Every business is an entity in itself. My business is my business. I can use any company's name to do it. Wherever I go, the networks go, and the business will follow." Wearing different hats under the same roof may also be used as a means of circumventing state regulations. After starting his body-shopping business, Piranavan registered a new company, Osin Associates International Pte. Ltd., in Singapore as a joint venture with Avol, a firm owned by a Telugu friend of his in Singapore. He used the name Avol to sign contracts with workers "on behalf of Osin Associates International" (see box 1 in chapter 6), but when placing workers he signed contracts with other agents in the name of Osin System Pty., registered in Australia. Workers suspected that Piranavan used multiple names, particularly those in Singapore, to evade regulations of the Australian authorities. Partly because of the practice of overlapping, Indian-run IT businesses in Sydney were characterized by an "over-registration" syndrome. Except for some résumé forwarders (see chapter 6) I interviewed, all the other businesses involved in body shopping were registered entities, often multiply. One informant estimated that there were more than 1,000 IT companies registered by Indians in Sydney and a more conservative estimate put the number between 300 and 400, but still, only 40 to 50 of these companies were active.

In sum, the body-shopping business developed through overlapping with other IT operations. Because of the feature of overlapping, for the entrepreneurial-aspiring IT workers arriving in Australia in the late 1990s, body shopping was not only a labor system bringing them to the global market as employees, but also a starting point for business development through which they could quickly accumulate their own resources in a transnational circuit, thus moving up from being transnational "technicians" to becoming transnational technopreneurs.

Chapter 5
Agent Chains and Benching

The fact that Indian IT consultancies moved to the body-shopping business due to the difficulties in marketing software products by no means implies that it was easy to establish direct connections with clients for body shopping. CSR Holding, owned by a Telugu couple—Chandary and his wife Shireesha, a psychologist by training—was one of most successful Indian consultancies in Sydney. After repeated but unsuccessful bids to become a labor supplier of large companies, they targeted state-government departments as clients instead, with the advantage of Shireesha's experience of working in various government agencies in Australia for ten years. Recruiting IT workers for government departments was less profitable, but also less competitive. CSR Holding successfully qualified to be issued the ITS881 status as a labor vendor for New South Wales state government. Following this, the couple soon made friends with government personnel officers and, subsequently, became a federal government vendor. This track record enabled CSR Holding to return to the commercial market and secure the position as a "niche skill supplier" for IBM Australia. IBM had two tiers of vendors: six big agents as its major vendors and a group of niche skill suppliers. When it needed workers, IBM first called in its major vendors; if not enough were supplied, IBM turned to the second tier of agents. CSR Holding had recruited only a few workers for IBM by the time I left Sydney in mid-2001, but the connection with IBM in itself gave CSR the cachet to attract good workers, both from India and in Sydney, whom they could place out to other clients or bigger agents.

The success of CSR Holding in securing clients, however, was an exception in the world of most body shops—only two body shops that I investigated had more than ten direct, stable clients. Instead they normally had to work with bigger agents to place workers. Piranavan, who ran a body-shopping operation under his consultancy Osin System, was forced to recognize this through an expensive lesson. He brought in five workers in November 1999, but could not even get IT companies to respond to his inquiries about openings. Piranavan finally placed his workers through other agents, and when I visited him again in May 2000, he was busy phoning all the big agents to identify the market trend. Joseph, a Kerala native, went to Australia in 1998 to set up a branch of a Chennai-based software service company for body-shopping business. Placing out only four workers in the first six months, he suggested the headquarters set up a unit in India to recruit workers for the Australian market separate from the one for the United States, because otherwise Australia would always be treated as "the second cousin" and only the workers who were not needed or were even rejected in the United States would be sent there (for the headquarters, the revenue generated in the United States per worker was much higher than that in Australia). When the proposal was turned down, he resigned and set up his E-Bet, determined to carve out a niche in Australia. As ambitious as he was, however, Joseph believed that dealing with clients directly was not where his "edge" lay. The realistic business focus, as he spelled out clearly, was to get in as a subvendor of big placement agents, as part of agent chains. Soul Networking, the biggest Indian-run agent in Sydney, brought in about seventy Indian workers over the two years of 1998–2000 and hired twelve recruitment consultants in early 2001.[1] Despite its respectable turnover, the manager believed that placing workers through other agents should remain the focus, even though this meant lower profit. This route was seen as the most efficient short cut into the market and to generate cash flow sufficient for quickly building up their infrastructure.

In order to secure their position in agent chains, body shops often had to do benching, as illustrated by how Kana started his body-shopping business in 1998. Kana was running an IT hardware trading business in Sydney when he came across a newspaper advertisement for two IT job vacancies just before leaving for India on a visit. In Hyderabad, his friends happened to have the requisite skills. Seeing the immediate opportunity for them to go to Australia, Kana contacted the company but was told that it would not sponsor overseas workers until they actually met them. Instead, they suggested that Kana use his own trading company as the sponsor, and should the friends prove satisfactory, the company would hire them as part of a joint project with Kana and pay their salaries through

71

him as well. Kana thus started his placement business and registered the IT consultancy Advance Technology. When he tried to expand the business, however, he immediately ran into problems:

> We know there is shortage for Oracle, SAP, etc. But I don't know who need them. . . . When I talked with companies, their first question is "where are the people?" or at least "where is the résumé?" What they want is—have people in their office; interview; decide instantly. We have to have people here. This is the first step.

Chaya had similar experiences when he started out in body shopping with his Sysway IT Consultancy in 1999. He spent the first six months calling up big IT companies in Australia and overseas. The telephone bill amounted to AUD 6,000, but no deal was clinched: "These big companies don't talk to you! I have to turn to other agents. . . . No agents will give you an 'order' to bring two or three workers for them. The way we do things is . . . ring agents up to tell them we have people here, and arrange the interview immediately."

Bringing in workers early and benching them was not always a body-shopping norm in Sydney. During the Y2K boom before 1999, big Australian placement agents took the initiative of contacting small Indian consultancies for workers; they even took care of all the paperwork, sponsorships, and "relocation fees," including airfare and, in some cases, accommodation in the first weeks. Indian consultancies hunted for workers and were paid commissions. As the Y2K boom ebbed, so did this relationship between big Australian recruiters and small Indian consultancies. In 1999 ICON asked Kana and Ravi to recruit six IT workers. ICON had sponsored their entry, but after interviewing the workers, all Indian, found only one acceptable and was prepared to foot the bill to send the rest back. Ravi and Kana decided to keep the five workers, who thus became the first to be benched at Advance Technology. For ICON, that was the end of such cooperation with small Indian-run consultancies; thereafter they only considered workers already on site. However, while the major Australian placement agents attributed the freeze in collaborations as due to body shops being an unreliable source of suitable workers, Ravi pointed out that the benching practice had produced a large labor reservoir in Sydney, which thus enabled them to recruit workers whenever they needed without having to maintain a relationship with any particular small agent.

Whatever the reason for the emergence of agent chains and benching, it is clear that the operation of the body-shopping business in Sydney was determined to a large extent by the role and positioning of big placement firms in relation to their corporate clients. IT labor recruitment in Sydney was dominated by the biggest five to seven placement agents, which in 2001 commanded 40–60 percent of the market share.[2] The extraordinarily strong market position of these big firms issued directly from their close, often intertwined business relationships with large corporate clients. From at least the 1990s, large companies had started to outsource their labor recruitment tasks to a limited few "preferred suppliers/vendors"—generally large placements agents. For example, Compaq Australia had reduced the number of its labor vendors from about one hundred to ten in the late 1990s. The largest placement agents in Australia usually occupied the role of preferred supplier for forty to fifty large clients, and 60–80 percent of the workers they recruited were placed in these companies. Around 1998, when the Y2K emergency brought about severe labor shortages, some large companies dispensed with their human-resource departments altogether, in effect outsourcing all labor-management tasks, including payroll and promotions, to a single placement agent—the master vendor.

The role of preferred suppliers and master vendors, once secured, meant that these placement agents' functions in labor management became deeply integrated into their clients' business development and were not easily replaceable. Most master vendors held formal monthly meetings with their clients, and kept themselves fully briefed on their clients' projects with frequent contact in between. Stephen Leo, a manager at Morgan and Banks Technology, described this relationship:

> We must keep communicating with the clients: "Hey, James, what's your development line for this year?" Then we advise them how feasible it is as far as the workforce is concerned. For example, we tell them that half of the workers must be recruited from overseas. They must be prepared for the cost. . . . Then we ask: "What is your timetable?" Based on the plans of all our clients, we will have our own time frame. We communicate with them again. Say, "Jack, would you put off your project for one month?" Or "would you put the project forward a bit?" That won't do any harm to them, but I can organize a good team [of workers] doing the same project for all the companies.

Large agents, thus, practically monopolized the access to large corporate clients such that smaller agents, even those who had perfect information and precisely matched candidates for a big company's needs, had to go through the preferred supplier or master vendor to fulfill a deal. Venu, a 457 visa holder from Andhra Pradesh, was intent on helping a friend migrate to Australia. His roommate Arun worked in ANZ (Australia and New Zealand Banking Group Ltd., one of the largest banks in Australia) and had been asked to organize a team of three short-term workers for a new project. Arun reserved one position for Venu's friend but could not bring him in directly because ANZ only recruited workers through its own vendors. Venu therefore approached a couple of body-shop operators, alerted them to the bank's new project and gave them the list of its preferred vendors. In return, he asked them to sponsor Venu's friend at a lower rate. Venu thus practically engineered an agent chain to bring in his friend.

Big IT placement agents' market position was also favored by the state regulation offering them special privileges as business sponsors of overseas workers. The Australian DIMA categorized sponsors of overseas workers into two types: standard business sponsors (SBSs) who were subject to a cap of twenty workers per petition, and pre-qualified business sponsors (PQBSs) who had no quota limits but had to first satisfy DIMA criteria with respect to size, business performance, financial standing, and demonstration of a continuous need for large numbers of overseas workers. Most big IT placement agents had had this pre-qualified status for years, which was renewed annually, and could thus bring in any number of workers at any time. Other big IT placement agents without a pre-qualified status signed labor agreements with DIMA that allowed them to bring in workers at any time so long as they did not exceed an overall limit set for the agreement period, usually three years; this arrangement exempted them from the annual reviews required by the pre-qualification scheme. On top of that, big placement agents might make special contributions to the government, hoping for favorable treatment such as swift processing. For example, for each worker it brought in, Mastech Australia donated AUD 500 toward the teaching of IT in Australian universities in 2000.[3]

Additionally, big placement firms had their own specialized immigration personnel who dealt with visa applications, liaised with government departments, and kept abreast of policy changes. Candle Australia Ltd., which managed an average of 1,500 contractors in Australia and New Zealand a year and attracted much attention when I was in Sydney due to its acquisition of the well-known IT recruitment specialists Unisys People and Alliance Recruitment, had an in-house immigration officer to supervise the applications from its different departments; and at ICON, a special team in Melbourne prepared all its 457 visa applications. As a result, a big agent required just about two weeks lead time to put an IT professional in place if the candidate was from western Europe or North America, and three to six weeks if from India. By comparison, the minimum time required for a body-shop operator to sponsor and bring in a 457 visa holder from India was at least one month. More than once, Glogo Consultancy founded by Samy from New Delhi ran into trouble because it could not bring workers to Australia in time. According to body-shop operators, the problem was not so much the slow procedure per se but that standard business sponsors, unlike pre-qualified sponsors, could not track the approval process sufficiently to anticipate changes and therefore make timely adjustments to their business deals. Thus, unless a body-shop operator secured a job opening after having workers on hand, it was unable to give the client a definite time frame for starting on the job. Furthermore, sponsoring a group of workers was more economical than single applications, partly explaining why workers were brought in collectively and put on the bench to be placed out individually.

Despite their privileged market positioning, big agents faced problems in recruitment, a major obstacle identified as "distance"—quite contrary to the oft-made prediction that IT would bring about the "death of geography" (Martin 1996; Cairncross 1997). Michael, an Australian placement manager, explained:

> Technology substantially narrowed down the physical gap, but the gap is still there. Placement is a very "high touch" business. . . . In a high-touch business, face-to-face communication is very important. Otherwise you can't have good judgments. When someone calls me to apply for a job, the only thing I say is "let's meet tomorrow morning." Nothing else. Everything is done through direct communication. . . . This is why the Indians can do well in the IT placement. They have the networks.

Partly for this reason, and to save the tedious tasks of dealing with large numbers of workers constantly, master vendors and preferred suppliers usually worked with smaller subvendors to recruit workers. For example, Hewlett-Packard Australia, which had previously outsourced its labor management tasks to about forty agents, in 2000 entrusted this to one master vendor, who in turn relied on eight subvendors for recruitment. Most of the staff that Candle Australia managed for its clients as their master vendor were recruited by its associate agents. Compaq's master vendor, Manpower,[4] worked with approximately fifteen associate agents, each specializing in certain skill areas.[5] Large placement agents also commonly paid a "spotter's fee" to the workers they managed who brought in persons whom they knew had the requisite skills.[6] When I visited Morgan and Banks Technology in Sydney in late 2000, I was handed a card reading "Sell Your Friends!" promising a spotter's fee of AUD 250. Some informants in big placement agents theorized that body shopping in Australia was an expansion of this common recruitment practice—in a sense, body-shop operators were full-time spotters.

Differentiated Circles

How were body shops connected to big placement agents within the agent chains? Here I use the examples of Piranavan's Osin System, and Joseph's E-Bet to map the relations. Piranavan was connected to four "circles" of agents, differentiated not by size but by how close they were to him. The outermost circle was kept open to all the agents he came across or contacted, and talking to these agents was Piranavan's main means of collecting market intelligence. The next circle consisted of the six or seven agents that Piranavan maintained relatively stable relations with and that he contacted for openings whenever he had workers at hand. The four Australian-run agents in the third circle were even closer to Piranavan; he was in their database, and they contacted him when they needed people. Piranavan's connections with these four agents arose in entirely different circumstances. One came about when the Australian agent hired a recruitment consultant from Mumbai who, it turned out, had a friend in common with Piranavan. Contact with the second came through an Indian contract consultant who had worked with the agent for more than six years. A recruitment manager of the third agent was dating Piranavan's neighbor and had been extraordinarily friendly whenever he came around at weekends; though the romantic relationship broke up, the business relationship with Piranavan continued. A cold call had brought Piranavan to the fourth agent, who was satisfied with the workers Osin provided, and this relationship was the strongest while the connection created through the Mumbai friend of a friend was the weakest, which to Piranavan was no surprise, as he always preferred strictly "professional relationships." Finally, Piranavan's fourth and innermost circle included three body shops, all bigger than Osin and run by friends, one of whom had found short-term projects for Piranavan's first group of benched workers at the end of 1999.

Joseph's E-Bet also worked with four circles of agents. The outermost was essentially the same as Piranavan's. The agents in the second outermost and third circles were similar to each other in both size and market niche, except that Joseph had worked with those in the third circle longer, followed their business closely, and, consequently, even developed personal relations with some of these agents' recruitment consultants. As a reflection of this difference in relations, E-Bet normally had to take the initiative to contact the agents in the second outermost circle for openings, while it was the agents in the third circle who contacted him for workers if they had openings. E-Bet's fourth, innermost circle comprised six body shops, all smaller outfits, and they had initially recruited one third to half of all the Indian workers that E-Bet managed in the year 2000–2001.

As with one of the connections between Piranavan's Osin and a big Australian agent, Indian IT contract consultants were sometimes in a position to pull the strings on a business deal as they often had long-term relationships with their Australian agents (one IT contract consultant often works with three or four placement agents only in searching for projects) and got to know the managers well. Mani Sandilya, a contract consultant in his late forties who sometimes worked on three jobs simultaneously, helped two body-shop operators place two workers through the big placement agent that he was affiliated with, and the body shops paid him a "finder's bonus" of AUD 500 for each worker. But quite often contract consultants helped body shops out of friendship. Puli Reddy, the owner of Puli Reddy Consultancy, regarded Indian contract consultants as an especially valuable asset: "These people are senior. Agents trust them. The companies [where they were placed to work] also regard them high. If *they* recommend workers to the agent or the company—very easy to go through."

The relation between body shops and big placement agents was not symmetrical. While body-shop operators approached as many big agents as they could, big agents were much less enthused about body shops. Candle Australia, for example, required any other agents who wanted to do business with it to sign an agreement limiting the maximum margin they would cut from workers' salaries. In a few cases where body-shop operators were found to be taking a bigger cut than agreed upon, Candle Australia had threatened to take over the workers' sponsorship to force compliance. Candle's worries were that workers paid too little might not be committed to their work, and that the clients would be unhappy if it was found out that they were paying so much to labor brokers. Some big agents only allowed managers with some authority to deal with body shops in order to impose better control.

Such stipulations to protect workers' interests, and tightened controls in dealings with body-shop operators, however, did little to help migrant IT workers, and only made the agents chain even longer, in the end causing workers to receive less of their pay. There was a trend after mid-2000 among body shops to recruit more workers from and through one another than to sponsor workers directly from India. This was partly because growing numbers of workers had been brought in and benched, particularly in the market slowdown, and partly due to big placement agents raising the barrier, forcing body shops to pass workers on among themselves in order to find a body shop that could in turn find a job for a particular worker. CSR Holding, for example, had not recruited any worker from other body shops in 1999, but by mid-2001, nearly one quarter of the workers it placed were provided by (smaller) body-shop operators. This leads to the question of the relations between body shops.

"Indians Are the Most Dangerous Ones!"

A mismatch between supply and demand was commonplace in the placement business, but this became particularly acute during the market downturn: more body shops had workers without openings while rejection rates of both big agents and clients increased. While immediate replacement had to be provided for the workers turned down at interviews or during trial periods, the rejected workers also had to be placed out as soon as possible to minimize losses. These tangles were mainly worked out through informal, spontaneous, and ad hoc connections between body shops. Collaboration among body shops of different sizes was more or less smooth. Remash and Keshev were college friends in Andhra Pradesh and flatmates in Sydney. Remash, an acquaintance of the founder and manager of Soul Networking, introduced Keshev to Soul Networking as a recruitment consultant. At the same time, Remash set up his own recruitment business, WinWin Recruiter. When I inquired about hiring a body-shop owner's flatmate without fearing that their clients would be hijacked, the manager of Soul Networking explained:

> Our payment [to Keshev] is based on how many workers he can place out. If Keshev does placement for WinWin, it's fine, as long as he doesn't do that in my office, using my fax or phone. The chance is very small [that] he can take our clients to WinWin. Our clients won't accept people from a new agent like WinWin. It takes a long time for them to develop the relationship.

What in fact happened was that within six months of Keshev starting his job, WinWin provided Soul Networking with five workers: Keshev's joining Soul Networking brought to it a new channel of recruitment.

Direct contacts between body shops of similar size was rare and often ended in disputes. The story of Rao Prabhala, who was working in the national railway company and at the same time running a body-shopping business in his spare time, provides a typical case. Rao sponsored a worker from India in 1999 and asked another body-shop operator, a good friend of his, to help find an opening. After the worker was placed, the friend proposed giving Rao a lump-sum commission and that he take over the worker's sponsorship. Rao agreed but was exceedingly unhappy. According to Rao, the friend was only entitled to a commission for finding the job; he should have remained the sponsor and kept the worker as a long-term profit source. Rao ceased contact with this friend. A couple of months later, Rao asked Mani Sandilya, the active body-shopping broker, to help place another worker. Mani introduced the worker to People Bank, a big Australian agent. Having learned a lesson, Rao wanted to sign the contract with People Bank directly, but Mani insisted that the worker's salary be sent to him first. With Rao and Mani both saying to each other, "I am not going to chase you for my money," the deal was canceled.

Knowing it was better to sidestep possible disputes of this kind, connections between body shops of similar sizes were often mediated through a go-between. Ravi was unique in the community in that brokering connections was almost his specialized business, capitalizing on his wide international connections. As a daily activity, Ravi contacted a long list of body shops in various countries to find out who had workers and who had openings, then tried to match them. When brokering a match, he involved himself as an independent party, taking receipt of the worker's salary from the body shop that had a client with an opening, and passing it on to the sponsoring body shop, retaining a certain margin for himself. In this way, the first body shop need not worry that the other would steal its client, and the sponsor was assured that the worker would not be poached. In one case, two parties who already had contacts with each other still invited Ravi to broker their collaboration because, as Ravi said, "they are not comfortable with each other, but are comfortable with me."

Ravi was also often called upon to settle disputes. In one instance, after he helped Puli Reddy place a worker through Greythorn, a large placement agent in Sydney headquartered in the United Kingdom, the worker changed his sponsorship to another body shop. Puli Reddy only discovered this after the fact, and turned to Ravi for help. Ravi asked the worker's new sponsor to pay Puli Reddy a large compensation, threatening to otherwise ask Greythorn to take over the worker's sponsorship and to muddy his reputation among other big agents. At the same time, Ravi asked the worker to cough up compensation as well. In talking about fairness in mediation, Ravi pointed out that a dispute could never be settled if the go-between was really neutral; he himself always took the side of the worker's original sponsor. This stance can be viewed against the regulation governing the Australian 457 visa that allowed workers to change their sponsors unconditionally, which many body shop operators saw as unreasonable and even impractical. If workers were free to do as they pleased, they argued, the risk of running a body-shopping business would be too high. Ravi's bias can thus be interpreted as a counter to the loophole in the regulation that could, ultimately, make the body-shopping business unsustainable.

Apart from doing business through go-betweens, connections between body shops were also made through links within the larger Indian community. After a Hindu religious lecture one Sunday,[7] Piranavan, always a keen attendant, mentioned to the day's speaker that he had workers on the bench. The speaker in turn mentioned this to Chaya, a friend of his, who had no previous contact with Piranavan. Chaya happened to know of an opening, and the deal between him and Piranavan was concluded smoothly. Piranavan volunteered that the cooperation would have been more difficult had it not happened through the speaker.

Overall, despite the need to stay in contact with one another from time to time, body-shop operators in Sydney did not work together closely, whether they were from the same region or caste in India or not. Fellow

Indians were often regarded as a potential threat: "Indians are the most dangerous ones," body-shop operators often said. This sensitivity was also obvious in other important transactions—with immigration solicitors and finance consultants. All body shops relied on professional immigration solicitors for going through the proper process, but I came across only two Indian lawyers who were retained by body shops on a regular basis. The majority used Australian lawyers. One Australian immigrant solicitor with an office in the Ashfield suburb worked for eleven Indian body shops, having gotten in touch with each agent individually, and all the agents had specifically asked him to keep any information about them from other Indians. Interestingly, Sri Lankan Tamil lawyers were regarded ideal choices as solicitors: they could be trusted to make certain adjustments to documents, say regarding a worker's skill level and remuneration (to make the application easier to approve), and relied upon not to leak this information to their competitors. Finance was another tricky area. On the one hand, body-shop operators wanted to show high revenues that might make their future sponsorship applications easier; on the other hand, they needed to underdeclare their profit in order to pay less tax.[8] Accounting was generally handled in-house, but finance advisors were sought on an ad hoc basis. Though there were many Indian finance advisors in Sydney (many more than Indian lawyers), the majority of body-shop operators consulted Australians only. Again, this was mainly due to the worry about business information being leaked to other body shops; some body shops also believed that finance records signed by an Australian accountant looked more credible and attracted less official suspicion than would an Indian name.

Overbooking Seats on the Bench

One direct outcome of placing workers through agent chains was "repeat placements," such as the case of three Tamil workers I came across who were placed out successively through five agents to a client. Agent chains sometimes ran so long and the relations between them became so complex that, one informant joked, body shops were themselves in need of "supply chain management" solutions—referring to the business software for optimizing order fulfillment and the movement of goods through the supply chain. In another typical example, a worker sponsored by a body shop that had made contact, through another body shop, with a big Australian placement agent was finally placed with a client. In this case, the client had a contract with the Australian agent, who was paid "service fees" in lieu of the worker's remuneration, as though the agent had assigned his own employee to work for the client. Big agents usually took a certain "dollar margin" cut as profit before forwarding the "service fee" to the body shop. (The dollar margin was a fixed-dollar-value deduction, in 2001, around AUD 10–15 per hour out of a fee charged at AUD 30–100 per hour.) The margin that the body shop in the middle took varied from case to case, but the general preference was for a percentage margin rather than for a dollar margin; in 2001 about 20–30 percent of the worker's pre-tax remuneration. If this body shop had close relations with the worker's body-shop sponsor, it might agree to a favorable (to the sponsor) dollar margin of AUD 5–15 per hour. The final deduction, the sponsor's profit, was not fixed. For workers with less than three years' experience, the margin could be as high as 50–60 percent (before tax), as was the case when the client paid a monthly "fee" of AUD 6,800 for a worker who ended up receiving only AUD 1,200! For workers with more than five years' work experience, the margin could drop to 20 percent.

As might be expected in unregulated transactions, body-shop sponsors sometimes calculated just how much, or how little, they had to pay workers in order to placate them, and simply kept the rest. Naturally, then, demanding workers would be paid more while more submissive ones might receive less. This was probably what Kana meant by telling me that in his business "psychology is more important than management." In some sense, the fact that the sponsor's margin was flexible made the "repeat placement" more acceptable to workers. This was why one worker had not attempted to change his 457 visa sponsor: "If one agent takes 40 percent, he may take only 20 percent when he makes a deal with another agent. They rely on us to make money. They must think about it."

Agent chains accounted for the often reported and puzzling phenomenon of a simultaneously high demand for IT labor on the one hand, and high level of short-term unemployment and low pay for migrant IT workers on the other. As an Indian H-1B holder in California was quoted as saying, the body-shopping operation could be likened to the way airlines overbook seats on flights—to always have workers left over in case they could not be sold.[9] Venkate interpreted the origins of the body-shopping practice in this way: "The [IT] companies need a big reserve of workers. But it is very expensive for one big agent to keep all the workers. Then these body shops came in. Each consultancy keeps couples of workers. Everyone can afford it." In sum, body-shopping operations and agent chains created a labor pool from which IT and other commercial corporations and state institutions alike could select and dispose of skilled workers anytime, yet without incurring costs for the state. The costs, of course, were picked up by the workers themselves.

Chapter 6
Compliant Bodies?

Benching and repeat placement raised two thorny issues for body-shop operators. First, how to prevent the workers whose visas they sponsored—their sources of profit—from running off upon finding themselves benched soon after arrival, an unpleasant experience for any migrant worker, more so for professionals who knew there was a market for their skills. Unlike what was common knowledge about H-1B visa holders in the United States,[1] the Australian 457 visa regulations allowed workers to switch sponsors unconditionally without repercussions on any future applications for PR status—in reality, therefore, workers could jump ship any time. Second, how to dissuade workers from lodging complaints with the authorities about unlawful body-shop practices, which could mean heavy penalties or immigration blacklisting for the body shop. The 457 visa sponsorship applications submitted by body shops were invariably generous with information that would meet the official criteria favoring the entry of highly qualified immigrants to help upgrade local skills. It would not be too long after their arrival that those whose visa applications were supported by, supposedly, attractively remunerated job offers would find out that this was hardly the case. For example, when Kana and Ravi recruited the six Indian IT workers from Singapore for ICON in 1999, promising each AUD 60,000–70,000 annually, the five workers who were found unsuitable by ICON were benched and told either that they could do casual work for Advance Technology, including answering phones and cleaning, for AUD 75 weekly, or be kept on a weekly stipend of AUD 25 instead. Under such circumstances, workers had more than enough cause for making an official complaint.

Yet, workers rarely asked to change their body-shop sponsor within the first year; as for complaints to the authorities, they were almost unheard of. Workers' explanations for this were simple—the fear of deportation. Body-shop operators were only too aware of workers' ignorance of their rights and did not miss the chance to reinforce this. Almost without exception, workers were deliberately misinformed that Australian laws required them to stay with their original sponsors in order to remain in Australia legally. I came across only four or five workers who suspected that this might not be entirely true but still wondered: "This is not possible in the U.S., how can it be so in Australia?" Thus, when workers expressed their discontent, they could simply be threatened with the cancellation of their sponsorship, or dismissed with the dare to "go take the next plane back to India." Some body-shop operators threw in intimidation, threatening to write to DIMA and damage the worker's future applications for PR status in Australia, or to defame the worker's reputation among major placement agents and IT companies, and so on. One well-established body shop told a worker who tried to change his sponsorship that it would make sure he would not be able to join any IT professional associations in the Asia Pacific region.

In order to cloak their control of workers with some credibility, most body-shop operators issued arriving workers with their own "agreements" or "mutual understandings"—which were completely different from the contracts submitted with their visa applications (for an example see boxes 1 and 2). Standard body-shop agreements put workers on a leash for a minimum of two years, and gave the sponsor the right to compensation amounting to three to five months of their "gross salary"[2] if workers left before that; some body-shop operators even required workers to surrender their passports and/or degree certificates. Agreements also forbade workers from approaching any clients or placement agents that they had come to know through their sponsors in the six months following the termination of their "agreement" period. Obviously, this was meant to prevent workers from directly hooking up with big agents, thus bypassing the entire agent chain and depriving body-shop operators of their cuts. All these patently unlawful and exploitative measures begged the question of why thousands of young, educated, ambitious, and even opportunistic IT professionals in Australia kept to the terms of their original body-shop sponsors.

Existing literature documenting labor-control measures adopted by temporary workers' recruitment agents (generally involving semi-skilled jobs in the service sector) has focused mainly on formal policies within the institutes, such as selective information transmission, employee recognition programs, maintenance of uncertainty (Gottfried 1991), a reliance on professional accreditations, or steep wage progressions (Cohen and Haberfeld

83

Contract

AVOL PTE LTD.

Dear T. K. Narendra:

We are pleased to offer you the position of Software Analyst Programmer in our partner company in Australia. This appointment is subject to the Company's terms and conditions of services in Appendix A.

Position:

Organisation: Osin
Reporting to: Manager, Osin
Location: Australia

Remuneration:

Your gross remuneration will be A$ 48000 per annum which includes Superannuation, Medical Insurance and Professional/Public Liability Insurance.

Benefits:

Salary review will be done at the end of 12 months. No additional contributions will be made to the Superannuation other than Government stipulated.

We shall be pleased to have the acceptance of this appointment within seven working days from the date of this letter. This offer will be withdrawn if we do not receive the confirmation within this period.

Confirm your understanding and acceptance of the above terms and conditions by signing and returning this letter to us.

We will be arranging for the Employment Work Permit 457 for you to work in Australia. Under this work permit 457 Visa, you are eligible to work only under Sponsorship of "Osin Associates International Pte Ltd" during the visa period. If the contract condition is changed during the visa period, Osin Associates International Pte Ltd shall take appropriate action as per the condition of employment explained and accepted by you in Appendix A. Please feel free to call us in Singapore at 65 _____ if you need any further information.

We wish to take this opportunity to welcome you to Osin Associates International Pte Ltd. And we look forward to working with you.

AVOL PTE LTD.
T. KUMAR
Director

I hereby accept your offer and terms and conditions as stated in this letter. I confirm I will start work on

Signed: _____

Date: _____   Name: _____

Box 1. Job contract letter from Piranavan's associate in Singapore on behalf of Piranavan. The absence of the Singapore addresses and the misspellings and grammatical errors are per the original.

Appendix

To:

Osin Associates International Pte Ltd.
(*postal address in Sydney*)

In consideration of Osin Associates International Pte Ltd (hereafter referred to as the Company) approving my appointment as Software Analyst Programmer to work in Sydney, Australia, I, T. K. Narendra, holder or Indian Passport No. hereby confirm and undertake that:

1. I will continue to serve the Company for a continuous period of TWO YEARS (hereby referred to as the contracted period) as from the date of appointment in Osin Associates International Pte Ltd, Sydney.

2. Shall I decide to resign from the Company before the completion of the full contracted period, I shall refund to the Company all costs incurred on me by the Company, subject to a maximum of 3 months Gross Salary.

3. I hereby confirm that I will not approach the company's clients, directly or indirectly for appointment with the client during the contract period with the Company and for six months after the contract is completed.

4. Shall I decide to join with the client, during the contracted period or within six months after the expiry of the contract period, I shall refund to the Company all the costs incurred on me by the Company, subject to a maximum of 2 months Gross Salary. In the event of you changing ther visa status (say Permanent Resident) during the contract period, you shall refund to the Company all the costs incurred for processing the work permit visa.

5. I agree that Company has the right to deduct the Tax at source as per the Australian tax rates as applicable and paid to the tax authorities on behalf of the employee. Also, after the actual tax is paid to the Government any excess amount deducted will be paid back to the employee.

6. I further agree that there are no implied responsibility for the Company other than what is stated in the appointment letter and the terms and condition in the Contract.

7. I shall not at any time during or after the employment with the Company, regardless of how such employment terminates, except as authorised by the Company or required by your duties, reveal to any person, the press and media generally or another Company or organisation, any trade secrets, or confidential information concerning the business or affairs of the Company or any of its related Companies or customers which may come to your knowledge by reason of your employment and shall keep with complete secrecy all confidential information entrusted to you.

8. Waiver: Failure or delay to exercise a power or right by a party to this agreement does not operate as a waiver of that power of right.

9. Severability: If any provision of this agreement is invalid or unenforceable it shall be severed and shall affect the validity of enforceablity of any other provision of this agreement.

10. Variation: This agreement shall not be varied in writing unless signed by the Employee and Representative of the Company.

Box 1. *continued*

11. Enter Agreement: This agreement is the entire agreement between the parties concerning the matters referred to in it and supersedes all prior contracts, agreements, understandings, negotiations and discussions concerning the matter referred to in it.

12. Governing Law: Subject to its terms, the Agreement is governed by, constructed and takes effect in accordance with the law of the jurisdiction in which the employee is based (New South Wales Law only, Australia).

I have read, understood, and accept the terms of this Agreement.

Dated at: This Sixteenth Day of June 2000.

Name: _____ Signature: _____

Witnessed by: _____

Name: _____ Signature: _____

Box 1. *continued*

1993; see also Kalleberg 2000, 349, for a literature review). Between body-shop operators and their stable of workers, however, labor relations were deeply embedded in a series of wider connections between body-shop operators and the larger Indian community. Moreover, the means for managing labor relations was largely informal and noninstitutionalized and had as much to do with the relations among workers under a particular body-shop sponsor, and workers' own strategies for furthering their longer-term goals.

Interlocks between Body Shops and Community Associations

Although most of my Indian informants took pride in the branded IT labor, very few held a positive view of the body-shopping business. While IT professionals tried to convince me that body shopping was only a transitory phenomenon and that the Indian IT industry would inevitably move up the global value chain, other Indians in Sydney, particularly the academics, condemned body shopping as unethical. But there did not seem to be any moves to sanction moral curbs on the practice; nor were efforts made to reach out to workers who were known to be in difficulties. In Sydney this situation partly resulted from the way the immigrant Indian community was structured, more especially the role played by IT professionals in more recent times.

Mutual Understandings

1. On arrival, you will be given accommodation and food/travel allowance of $100 per week until such time you will be placed on Contract for professional placement. This is applicable only to employees agreed upon earlier.

2. From time to time you will be given training on local communication and work practices from the Company or any of our Associates.

3. Whenever possible, the Local Manager from Osin Associates International Pte Ltd will brief you and assist you to get you placed as soon as possible.

4. In the past we have placed people within three weeks of arrival subjected to display of your professional skills and communication skills.

5. We are very keen to place your professional skills at the earliest possible time to achieve maximum benefit to both the parties involved and shall fully adhere to local rules and regulations in full.

6. The Government delay in processing your Visa application on arrival is not in our control and we shall ensure to get them approved as early as possible.

7. The Employee's Passport will be safely kept by the Company in a common place (Accountant or Lawyers offices for security reasons).

You will be reporting on a regular basis to Manager, Sydney on your work to interact with other staff.

Box 2. "Mutual Understandings" issued by Piranavan to workers on their arrival in Sydney

At one time, it seems, one almost could not expect to become a leader of an Indian community organization in Australia in general without being a medical doctor. But from the 1980s, and particularly the early 1990s onward, community leadership positions were increasingly taken up by the more recently arriving professionals, mostly with engineering backgrounds. By then, too, this constituency formed the majority in the established community associations, or set up new associations often due to conflicts with the earlier arrivals. A consequent strong interlocking of interests between community association leaders and body shops was difficult to miss. Among the thirty to forty entrepreneurs and partners who ran the thirteen body shops that I had close contacts with, at least one third were key organizing members of various ethnic organizations (my estimates based on the membership lists I collected), and many more were actively participating members. These dual affiliations were also quite evident at Indian ethnic functions that were used as occasions for advertising body-shop businesses. For example, Chaya was the president of an Indian Tamil association in Sydney and, at the same time, a body-shop operator, so one often found announcements such as "urgently needed: embedded C programmers" together with "scintillating lectures on Bhagavad Gita" within the same e-mail he sent out to association members and business associates. Nadarajah, Chaya's business partner for their body shop cum training institute TrueInfo, rushed from one suburb to another to deliver IT courses after giving a religious lecture every Sunday, accompanied by quite a few students also running from Hindu discourse to Java software language. Nor was it unusual for association members to approach bodyshop operators at ethnic functions for help in sponsoring relatives or acquaintances from India.

Underscoring this transition in the leadership and composition of the Indian community in Sydney was the marked shift in the focus and nature of community associations—from previously politically tinged concerns to "culture." Where earlier associations were keen to interact with the mainstream society in Australia, foster links with government in India, and represent the entire Indian nation, from the 1990s, these initiatives were displaced by far more strenuous efforts devoted to cultural and religious activities, and to forming regional associations organized along Indian states of origin or language groups. The overlapping religious and cultural involvement of body-shop operators not only elevated their social status but also appeared to win them genuine respect from workers. Piranavan, always in very fine kurta, held two big *puja* rituals at his house each year, to which he invited all his workers. The more people attending a *puja* the better, I was told (and invited); and when I was unable to go on one occasion, Susai and Suman, two young workers benched under Piranavan's sponsorship, described it to me in great detail the next day and insisted on sharing with me the *prasadam* (ritual foods handed out by the priest during *puja*). "This is food from God, you must have it," said Suman. "I was thrilled. I didn't know Piranavan was so religious." After that Suman started doing a *puja* every morning. What impressed the workers was that body-shop sponsors were deeply religious while at the same time successful and aggressive in business. (Merely being religious without having a successful career meant little, while those who did well in business but were indifferent to religion were viewed skeptically.)

Concomitant with the leadership transition was a reformulation of the roles of those in charge of community associations, which cast them as "organizers" rather than "leaders." While the authority of a "leader" is based on power, resources, or personal charisma, that of an "organizer" relies more on having wide connections and actual performance. This appeared to closely mirror the changed orientation of community associations from the political to the cultural, and from the national to the regional. As organizers of regionally oriented cultural associations, bodyshop operators thus developed wide and deep connections within the community (most association members whom I met commented that the organizations were now more "democratic" and "people oriented" than before)—and this created a friendly and protective environment for their business. How could a regular participant of the events organized by the Tamil association presided by Chaya offer to help any worker Chaya sponsored to challenge his authority? Kalaimani, a full-time IT contract consultant and a part-time *jyotishudu* (palmist), sometimes passed on job information to the unemployed or benched workers who came to him for personal predictions. Kalaimani was known throughout the community for his kindness, and once, after learning that Uday and I had been living without a refrigerator, he took the trouble to transport an old one from a relatives' place to our flat. But, sympathetic as he was, he would always discourage workers from taking any action against their body-shop sponsors and was careful not to be seen as an instigator, emphasizing instead divine blessings for agitated workers on the bench.

It follows, then, that anyone accidentally caught in a confrontation between body-shop sponsors and workers would more likely step in to pacify the worker, or at least not undermine the sponsor's interests. Venush and Umesh, two of the first five workers that Piranavan brought to Sydney in 1999, were desperately anxious while they were benched and took to personally calling on Piranavan to ask about jobs. I was in Piranavan's sitting room cum office on one occasion. Piranavan swiftly led them into the kitchen; about ten minutes later, he emerged, showing them to the door. But the two did not leave and seemed to have more to say. Also in the sitting-room office were two middle-aged Indians, a man and a woman, working at computers (I later learned that they were doing part-time work for Piranavan). The man rose to the situation: "What are your platforms [skills]? . . . Very good! I don't see any problems. Take it easy, young men." The woman also chimed some reassurance. Venush and Umesh had no choice but to say good-bye and leave. This sort of supportive intervention was not seen as "helping" a body-shop sponsor at all; instead, informants argued, the only "constructive" thing a third party could do was to minimize such conflicts and encourage workers' obedience—because it made the most sense: "What can you achieve by being confrontational?"

Compared to their body-shop sponsors, young IT workers had very limited social circles and were cut off from the larger immigrant community. In most cases, they socialized only with others under the same sponsor, and more rarely mixed with non-IT Indians. One worker attributed this to the fact that the non-IT Indians usually worked out of the downtown area so it was difficult to meet them for lunch or run into them on the train. (Most IT companies, or companies with large IT departments such as banks, were located in the two central business districts [CBDs] in central and north Sydney, while conventional industries including manufacturing were usually in the west.) The bachelor status of most IT workers also inhibited wider mixing in the community where reciprocal dinners, wives' shopping together, and children's groups were the central avenues fostering close relationships. Bachelors were reluctant to invite families to their shabby flats shared with five or six men, let alone reciprocate dinners. A particularly sharp line was drawn between bachelors and families, and intimate interactions were sometimes seen as strange, even suspicious. Srina and Shyla, the couple living in the building next to Uday's while I was there, cited their regular contact with us as proof of their self-claimed open-mindedness. In short, young IT workers' isolation prevented them from seeking support from the wider community to counter the control and surveillance of their body-shop sponsors.

Workers as Intermediaries

Depending as they did on associates and personal networks for recruitment and placement, the business of body shopping naturally involved a large number of middlepersons. But even here, go-betweens who brought a body-shop sponsor and a worker together rarely intervened on the worker's behalf. Two out of the five workers that Piranavan brought in from India in November 1999, Rajan and Jathavi, were recruited through a common friend. Rajan had been a lecturer in an engineering college in Coimbatore, a town in western Tamil Nadu bordering Kerala and another large base for the production of IT professionals, and Jathavi had been a student of the same college. Piranavan got in touch with them through Lahudoss, his family friend, who was also Rajan's colleague and Jathavi's teacher. Rajan and Jathavi contacted Lahudoss when they found themselves benched. Lahudoss told them that the market was bad during the Christmas and New Year holiday season and assured them that everything would be fine soon. This was exactly what Piranavan had told them, and, most likely, Lahudoss consulted Piranavan before replying to Rajan and Jathavi. The other three workers, Venush, Umesh and Susai, were introduced to Piranavan by a body shop in Hyderabad. They too had tried to contact the body shop in Hyderabad after being benched for more than two months, but got no reply. In another case, Govindan, who had more than ten years' work experience in IT, was left in a plight for a few months with his pregnant wife after they went to Australia via Jamaica under the sponsorship of World Digital, a body shop run by David from Kerala. David had promised Govindan an annual salary of AUD 120,000 plus free accommodations for the first two months, but when he arrived, David told him at the airport that there was no job and dropped the couple off at the two-bedroom flat rented by Uday—to become the fifth and sixth tenants. When Govindan turned to the former college classmate who had introduced him to David, he was told to be patient. Although college classmates are among the closest social relationships in urban Indian society, Govindan doubted that his classmate would raise the issue with David.

Nor did workers expect go-betweens or middlepersons to help them out; some even felt that having a friend in common obliged them to cooperate with the body-shop sponsor. Sam, an IT worker in his thirties, was introduced to Chaya by Chitra, their common friend. Being on the bench for two months without pay, Sam complained to me reluctantly but repeatedly that Chaya was "mean," but he did not blame anyone: "I am Chitra's friend. Chitra is Chaya's friend. Chitra told me that I might have to wait for a while to find a job. He also told me that the agreement [formal contract] might not be 100 percent true. But if we had not made up one, I couldn't have gotten my visa in the first place. . . . Things are worse than I thought. But what can I do? I don't blame them."

To interfere in "a formal business relationship" (between body shops and workers) on the basis of a personal relationship was, I was told, "unprofessional." One informant asked: "If I helped arrange a marriage, should I barge in when the couple fight?" Sharma, a middle-aged Telugu engineer, introduced his brother-in-law in India to Samy, the owner of Glogo Consultancy. The brother-in-law was unhappy with the job he was given and wanted Sharma to ask Samy to send him to another company. Sharma complained to me about getting caught up in family politics. With a pained expression he repeated to me how he had replied: "Everything has a limit. You are not supposed to go beyond the limit. . . . Samy helped you come here. How can I be so greedy and push him all the time?" Sharma did not speak with his brother-in-law for a while after this.[3]

Many intermediaries were reluctant to take the workers' side also due to their own business interests. Govindan had grounds to doubt that his classmate would speak to David on his behalf because the classmate— also a body-shop operator—wanted David to become his business associate. This point can be clearly illustrated by the role of a special group of intermediaries who can be termed "résumé forwarders," and comprised workers once managed by body shops, or even still under their sponsorship. All body-shop operators had a network of such workers who helped their own contacts migrate to Australia or find different jobs and/or sponsors in Australia by forwarding their résumés for a fee, often as a run-up to starting their own body-shopping business in the future. Résumé forwarders had basically two types of clients—those with relevant work experience, and those without who wanted to buy visas and land jobs on their own once overseas. Understandably, the second group was the main profit source and was charged as much as AUD 6,000–10,000 for a 457 visa in 2000, from which the forwarder took anything between AUD 1,000 and 2,000 (the charge for a 456 visa was much lower, normally less than

91

AUD 1,000). When résumé forwarders brokered the sale of visa sponsorships to workers without proper IT skills, they had to guard against these workers turning the tables on the body-shop sponsors and holding them to the terms of the job contract submitted to the immigration authorities. Uday was as an active résumé forwarder for David, but never worried that his clients might give either him or David any trouble: "They know they have to find jobs themselves. They just follow whatever I told them to do. If I don't tell them they can go to court with the [contract] letter, how do they know?" For workers with IT skills, body-shop operators sometimes paid the forwarder a commission of AUD 500–2,000 six months after the worker was placed; hence, the résumé forwarder would get nothing if he failed to pressure the worker to stay with the sponsor for that period.

Résumé forwarders sometimes even helped body shops in policing workers with whom they had no relationship. After David left Govindan and his wife at Uday's flat, in the beginning, Uday looked after them well; he showed them around, brought them to Hindu temples and Indian groceries, and referred Govindan to Kalaimani to have his palm read (without being charged the usual AUD 10 fee since he was introduced by Uday). Uday also introduced Govindan's wife, an IT professional but with no work visa, to the body shop G&J IT Solutions run by a father and son, Rajiv and Kejal, from Mumbai. However, when Govindan went ahead to change his own visa sponsorship to G&J, Uday turned nasty. He accused Govindan of damaging his relations with David. When Govindan moved out of the flat, Uday asked for the rent, and when Govindan insisted that David pay, Uday slammed the door in his face. Govindan then told Uday that he would pay once he had enough money. But Uday was unhappy that Govindan was delaying payment and wanted to file a police complaint against him: "The police must support the local people. I am a PR. They are only temporary workers. When they have this kind of record, they can't apply for PR [status]."

Relations among Workers: Support Yes, Solidarity No

Mainly due to a reliance on personal networks, body-shop operators normally only recruited workers from limited places in India. Does this mean, then, that workers could easily form alliances based on their place of common origin to keep body shops in check? In reality, the connections between workers tended to promote, rather than undermine, the body-shopping business. The Indian IT migrant workers had strong support networks among themselves, and mutual assistance helped in mobilizing resources to go overseas and supported unemployed workers through hard times in the destination countries. Before Ravinder moved to Australia on a 456 visa under Puli Reddy's sponsorship in 2000, two of his former classmates in the United States sent him USD 500 each to pay off Puli Reddy's body-shopping fee and cover his airfare. Ravinder had less than AUD 300 in his pocket upon arrival in Australia; Uday shouldered all his expenses while he stayed at the flat. When I moved out temporarily, Uday asked: "How much can you help me?" My share of the rent for the three weeks I stayed came to AUD 100, but Uday was asking for AUD 250. According to him, since he was the only person in the house working, his brother Ashok and Ravinder were part of his responsibility, but as I was clearly not I was subject to overcharge to share his burden.

More drama followed. Immediately after I moved out of the flat in December 2000, Uday found AUD 3,000 missing from his credit-card account. Uday promptly made a police report identifying me as the prime suspect because—as he explained later—I was on a visitors visa (one of the first things that Uday had inquired about when we got to know each other at Advance Technology) and because I had told him that I was not interested in becoming an Australian resident. Therefore, Uday told me, he felt that, unlike most Indian IT workers eager for a foothold in Australia and anxious to not do anything wrong to spoil the plan, I could "do anything"! Fortunately, just before the police tracked me down, Ravinder confessed to giving Uday's personal identification number (PIN) to Puli Reddy. It turned out that Puli Reddy had threatened to not convert Ravinder's 456 visa to a 457 unless he paid up AUD 3,000 before the end of 2000. Ravinder found no other choice than to "steal" the PIN and pass it on to Puli Reddy! As money conscious as Uday was, he did not blame Ravinder but regarded the AUD 3,000 as a loan.

Sometimes networks of assistance expanded quickly. Having nothing to do in Sydney, Ravinder visited Melbourne and stayed with a friend of his former classmate's for three weeks. The friend offered to lend him AUD 300, taking this from his second month's salary as a railway-station security guard. This kind of friendship was taken for granted; as Rajan pointed out: "There are so many young Indian IT guys in America. How can they survive there if there are no such friends? . . . Western people have no sense of community. Their 'mates' are not real 'mates.'" Hence, some informants told me, local workers in the United States were more reluctant than Indian workers were to move to the highly expensive San Francisco Bay area: Indians could cut their living costs by sharing flats but Americans could not. Govindan explained that Indians in the United States usually got married as soon as they got H-1B visas so that they would not face endless requests for financial help: "You can't ask [for] money back after lending to them [friends or relatives]. Because you are in the *U.S.!* You have the reputation. . . . If you are married, your friends can't ask money from you."

Workers' support networks did not, however, lead to either solidarity among workers or to winning them bargaining power against body-shop sponsors. A typical example of the lack of solidarity and collective action is how Uday and Sree Kumar, a young IT worker from Kerala who was also sponsored by Kana, handled pressure from Kana. Uday and Sree Kumar were very close friends, and Uday was particularly grateful to Sree Kumar for introducing him to David when they were both on the bench, which was how Uday subsequently got a job at TNT, a giant business-to-business express-delivery company originating in the Netherlands. As soon as the news of both Sree Kumar's and Uday's getting TNT jobs reached Kana, Kana talked to Uday about "compensation." Uday agreed to Kana's demand for AUD 1,000 monthly for as long as he kept his job at TNT. Uday believed that Kana must have asked Sree Kumar for compensation as well but, surprisingly, the duo never discussed this matter. When I persisted in asking why they had not done so, or tried to bargain for lower compensation each, Uday shot back:

> Isn't it clear? Kana wanted money from us! Money is a bloody thing. Money spoils so many relationships. I never asked Sree Kumar how much he made. If you earn money and you use it yourself, it is all right. If other people are involved, there are always problems. [*XB: But if you two had talked about that, you might have worked out a number, for example five hundred dollars. If you'd paid the same amount to Kana, it would have saved a lot for both of you.*] How can we work out how much we should pay? If we pay the same to Kana, Sree Kumar will feel bad. He has a family to look after and I am single. If I pay more than he, I am unhappy. . . . You are right, I might have paid less if I talked with Sree Kumar. But I am happy to pay the five hundred dollars more. I don't want to talk about money with Sree Kumar. I don't want our friendship to be damaged.

The highly flexible labor market engendered very uncertain and unclear working situations: workers did not know when they could get a job, for how long the job would last, how much they would be paid, and how much they should ask for. Since salary was supposedly determined according to the worker's merit as well as the ever fluid market rates, workers found it difficult to discuss their individual situations between themselves, let alone to take collective action. Negotiations between workers and body-shop sponsors were always conducted on a one-to-one basis and a worker's net pay was treated as a business secret. Asking about or demanding more information on how salaries (even one's own) were determined was "unprofessional." Ironically, thus, upholding professionalism here served to undermine rather than to enhance transparency.

The lack of transparency was compounded by an idiosyncrasy peculiar to young Indian IT workers. Workers were so wary of attracting the envy of fellow workers that they kept their intentions and life plans, including the problems they faced, as closely guarded secrets.[4] When Uday applied for PR status, for example, he did not even tell his cousin who was visiting Sydney: "If I tell this to someone, I will fear something may happen. We feel that the person may write letters to DIMA saying I have a bad character." And when Uday's brother, Ashok, landed a job with Microsoft Australia and was being sent to the United States for training, Uday's other flatmates were told that he was going to the United States to look for jobs: "We can't trust these people. If they know the truth, they may pray to God to destroy our prosperity." These sensitivities were so strong that I had difficulties in employing the standard snowballing technique in my fieldwork; not only did people hesitate to introduce other workers to me, but, more than once, informants called me after interviews to reconfirm my assurance that everything they told me would remain absolutely confidential. Although this sensitivity about money, or the need for secrecy in general, appeared to be a cultural issue, it gained special significance when situated in a particular institutional context. Uncertain work conditions on one hand created secrets for each worker and, on the other hand, generated an atmosphere of fear.

Solidarity among workers was further undermined by their internal differentiation. Not that the workers were fragmented with conflicting interest, but workers who were the first to find jobs through the body shop, or who were directly employed by the body-shop sponsor, often took the initiative to smooth over relations between benched workers and the body-shop operator. The relationship among Piranavan's workers again provides an illuminating example. Jathavi, Venush, and Umesh were finally placed in May 2000, almost half a year after their arrival. This was a boost to business, and in July, Piranavan rented a small office. Rajan and Susai worked on software-development projects in that office. Jathavi, Venush, and Umesh were all paid AUD 2,200 a month; Rajan was paid 2,500; and Susai 1,000 only. Umesh estimated that the client paid Piranavan AUD 4,000–5,000 for him, but he saw nothing to complain about because Rajan, the most talented among them, was being paid a similar salary and it would be too arrogant, Umesh said, if he demanded to be paid more. When I asked Rajan why he did not shift to a bigger agent to seek better pay, he responded, in his usual unhurried and soft way, "How do you know the new agent will be better? . . . I came here through friends anyway. How can you trust a stranger agent so easily?"

Piranavan's second group of workers, three altogether, arrived in September 2000. Piranavan asked Rajan to pick them up at the airport and help them settle in. They too were put on the bench. One of them, Kasi, a skinny but hyperenergetic man from Coimbatore, lost his patience. When I visited him one Friday night he told me that he had had a "fight" with Piranavan that day: "I don't care he is [an Australian] citizen or what. I talked to him about my difficulties, he spoke English to me! Bullshit!" (Replying in English to someone speaking Tamil or Telugu was seen as being deliberately indifferent or rude.) The "fight" had actually lasted only five minutes, as Piranavan, rushing out for a meeting, had pushed Kasi on to Rajan. "He [Kasi] thinks doing business in Australia is very easy," Rajan recounted to me, "It isn't. I know that. I told him." When Rajan visited Chennai at the end of 2000, he brought back four shirts for Kasi from Kasi's relatives. What could Kasi say, no matter how disgruntled with Piranavan, when Rajan chided him to look at the positive side of life in his soft tone?

At Kana's Advance Technology, Vand played a role similar to Rajan's. Vand told me proudly that he always intervened when there was a conflict between Kana and the workers: "This is good for the company. They listen. Whatever I say to the boss, the boss always says yes." Once, Kana had complained that Arun, a worker at Advance who was placed to work in the ANZ Bank for a while, was lazy. Vand agreed, but in the next month did a large part of Arun's work without telling Kana. From November 2000 through February 2001, none of the software-development workers at Advance was paid, because Kana could not secure any business deal, and all four of Vand's flatmates practically lived off his savings.

The workers who served as a buffer were often unusually competent, and the strongly held assumptions of "meritocracy" among workers were intrinsic to their compliant role. For example Vand was appointed director of Kana's software-development outfit and paid AUD 6,000 monthly, which was rare for those working for small firms. When unemployed workers complained, body-shop operators often shut them up by pointing to these talented middle-level workers: "Can *you* be as competent as them?" Vand thought the body shopping system was perfectly fair:

> In these eight months many things happened. When I just came, I was not paid one paisa. After two months, the salary doubled. Now it is even higher. . . . Why did you say there is fraud or cheating in the system? If you can prove that you are good, you can always have a good salary because the company *knows* that they have to rely on you.

Ravichandary, the only son of the personal assistant to a state governor in India, had paid Kana AUD 8,000 to get his 457 visa. Despite his family background, Ravichandary worked as a receptionist at Advance Technology for AUD 800 a month. He was depressed, but not complaining: "I am forty years old, how can I compete with these young people?" he said, pointing to Vand, who sat in a room of his own. Ravichandary had a master of arts degree from a decent university—but this was no better than being "illiterate" [*sic*], as Uday tagged him.

The Way Out

Workers were not always willing accomplices in their sponsor's business plans, and a fuller picture of the labor management mechanisms in body shopping must take account of these ambitious migrant workers' strategies. Hirschman (1970) summarized the basic responses of people dissatisfied with an organization in a market economy—"exit," "voice," and "loyalty," which activates voice and reduces exit.[5] In body-shopping arrangements, "loyalty" is seen as completely outdated by both workers and their sponsors, and "voice," or expressions of any kind of protest, are firmly discouraged. Workers' preferred strategies, namely making a "quiet exit" or, more commonly, taking the "escalator," can perhaps be seen as subcategories of exit. The quiet exit meant secretly changing sponsor, while the escalator exit meant leaving a body shop (openly) after having (secretly) obtained Australian PR status or a visa to other countries, particularly the United States, even before the "agreement" period was up. These strategies did not correct lapses of the system, but lent body shopping a self-sustaining mechanism.

Why did the workers give up their voice? This question becomes even more intriguing when one considers the relatively pro-employee labor regulation in Australia in general. My informants pointed to "fear" as the reason for not fingering their sponsor. But fear is far from a simple psychological reaction; one would have to be socially informed or coerced to be aware of the potential danger of certain calculations. The experiences of the only three workers whom I came across who threatened to sue their body-shop sponsors illustrates how the fear or caution was socially produced. Sudheer, an IT professional in his late forties, was recruited by Kana from Singapore in 1999, together with Uday. But when Sudheer was put on the bench, he threatened to raise a complaint unless Kana paid him the full salary for this waiting period as per the contract submitted to DIMA. Kana relented and paid him the named sum of AUD 6,000. Sudheer was an exceptional case because he went to Singapore, then to Australia, while still on medical leave from his employer in India with the sole aim of making quick money for his daughters' dowries. His voice strategy—successful because he was seen as desperate—was thought hotheaded and even silly by his fellow workers. About one year later, Sudheer contacted Uday to see whether he could go to Australia again; Uday said he would look into it but told me that he had no intention of taking it seriously: "If the consultancy knows what Sudheer did, no one will sponsor him! You can earn the six thousand dollars in two months here."

In the second case, a worker who went to Australia with Piranavan's sponsorship in early 2001 refused to report to the office and, instead, sent back the "Mutual Understandings." Piranavan warned him that he would cancel his sponsorship unless he cooperated, to which the worker's riposte was that he would sue immediately if Piranavan did so. Piranavan stopped right there. This worker, also thought strange by most of his fellow workers, was in touch with no one. Some commended him as a man with teeth, but just as quickly added that it was a matter of personality and could not be simply emulated.

The third individual to challenge a body-shop sponsor was Govindan. A couple of weeks after his arrival, David introduced him to an Indian-run company with four workers for a job interview before rushing off to Singapore and the United States on business. Govindan, who used to be a team leader supervising forty workers in India, turned the job down. He asked for my advice. I said that he could take David to court, which he ruled out immediately: "This is no good for anyone." When David returned two weeks later, he made clear to Govindan that he had only two options: take the job offered or continue waiting on the bench. Govindan accepted the job. The Indian company then decided to pay Govindan half the salary promised at his interview. Govindan was outraged; in December 2000 he changed his visa sponsorship to G&J. From December to March 2001, David made four calls and wrote three letters asking Govindan for a compensation sum of AUD 25,000. David also claimed that he had registered a case in the Institute of Arbitrators and Mediators and was ready to make a big fuss. Govindan found a lawyer, made me promise to be a witness in court if required, and also asked me to look for a suitable journalist, just in case. However, all these were merely preparations for the worst possible outcome, as Govindan had made up this mind not to take any action against David so long as David did not proceed any further either. Just thinking of the legal battle, according to his wife who was now seven months pregnant, had been a torture for them. Indeed, for workers who had minimal support in Australia and just wanted to better their lives, why should they be eager to raise their voice?

What Uday and Sree Kumar did, getting jobs at TNT through David without their sponsor Kana's knowledge, is a typical case of the quiet exit. Apart from avoiding a confrontation with Kana, Uday had another reason for being quiet. He actually believed that a change of sponsor would handicap his future application for PR status in Australia: "The immigration department will ask 'why did you change?' If I were the immigration officer, I won't process the application. Naturally we think this person is not stable!" Venkate, formerly a salesman for one of India's largest insurance companies, was one of the most sophisticated IT workers I came across. When he was benched by Advance Technology, he read the newspapers every day and visited immigration solicitors, finance consultants, and lawyers to figure out how he could become a PR as soon as possible. He lodged his application for PR status soon afterward and obtained it nine months later. He then told Kana and other workers that he had to leave for India, as his mother was unwell. It was another two months before he told his Indian friends that he had already been an Australian PR for months. Vand, despite his commitment to Advance, was also exploring the possibility of a quiet exit. He entrusted Uday with the task of finding him jobs in big companies. According to Uday, I was the third person in the world who knew this "top secret." In sum, the strategy of quiet exit opened up a way for workers to extricate themselves from difficult situations without challenging the body-shopping business in any sense.

But only a few workers had a secure alternative enabling them to brave a quiet exit; the majority hoped for an escalator ride. When a worker left a body shop having become an Australian PR, or having secured an opportunity in another country, the body-shop operator was normally quite happy to let the worker go. Usually, leaving for an opportunity in the United States was treated more favorably than having secretly become an Australian PR: while the latter workers were allowed to terminate their agreements only a couple of months early, those going to the United States were let go much earlier without having to pay compensation. Prakash, a worker at Advance who was said to be the second most talented IT professional after Vand, left for New Jersey on an H-1B visa in March 2001. He had been a real asset for Kana: his skills were worth AUD 100 an hour but Kana paid him only AUD 15. Kana did not object to Prakash's departure and even proposed a farewell party, though that did not happen. Many body-shop operators recounted to me with some pride how many of their former workers were now in the United States or had become Australian residents. Indeed, the workers' hopes for moving up and, particularly, becoming IT entrepreneurs one day, were a key reason for their compliance with body-shopping practices. The strong entrepreneurial bent among the IT professionals also made them regard uncertainty as a natural business pressure that body shops could neither control nor be blamed for. Workers preferred the escalator exit over the quiet exit also because, while a quiet exit would almost certainly lead to tension between the body-shop-sponsor and the worker later, escaping by escalator sometimes brought about a close relationship with the sponsor, which facilitated the (former) worker's push to become an entrepreneur, particularly by acting as a résumé forwarder.

Chapter 7
The World System of Body Shopping

Venugopal, a twenty-four-year-old Kamma Telugu who went to Australia sponsored by David in 1999, had a three-step career plan. In Australia, he would accumulate as many connections and as much work experience as he could. Hence, twice a week, instead of the home-packed lunch to save money (as most fellow Indians did), Venugopal made a point of lunching at his company's café in order to meet more people. His next step, after two years, was to go to the United States: "After working two years in a Western country, you can understand the entire work system. Like learning cooking: after two years, you can cook any meal on the menu." But why, I asked, was it still necessary to go to the United States if one could already cook anything on the menu? "To do business, you must know about everything. In the U.S., you will know about the *world* market. Then you can make the right decision. . . . I miss my friends [there] as well! We must meet up to sort things out: who will go back to India, who will stay in America, who should go to other countries." This was Venugopal's career goal—step three: to set up a globally operated business in either the United States or India.

When I got in touch with Venugopal again five years later (2005), he was, however, still in Australia. With an apartment purchased and a marriage negotiation just concluded in Andhra, he planed to get married in October in India and then bring his wife to Australia to settle down. It was only at this time that I found out he had lodged his application for permanent residency in Australia while he was convincing me of all the absolute necessities of going to the United States. He was now planning to have his honeymoon in the United States, since he had "so many friends there, [it] must be fun."

Venugopal had not lied to me. Indian IT workers in Sydney were just as eager to leave for the United States as to become Australian PRs. Among the IT workers in Sydney, there was no escaping the ramifications on social identity of one's immigrant status, which inevitably cropped up as a topic of some interest (or sensitivity). When I first read out the names found in Uday's address book, he immediately qualified each one with information on their visa or residence status, pending or otherwise, without even pausing to recollect. Shyla, an occasional visitor to our flat who lived in the next building, was not acquainted with any other Indians in Uday's building but still knew which one had recently become a PR. When I first met Siva from Vishakhapatnam, Andhra Pradesh, he was in a state of anxiety: his father was very ill and the family wanted him to marry as soon as possible but his Australian PR status was still pending. Though Siva's master's degree in IT from Queensland University was imminent, with no confirmed PR status in hand he would have difficulty landing a good dowry: "In India, people have no information about Australia. If we tell them that IT students can get PR [status] easily, people won't believe it."

Australian PR status was valuable precisely because it brought not just the right to settle down, but also the convenience of moving on and having a base to return to. As Venkate had responded when I belatedly congratulated him on becoming a PR: "Yes, I can go anywhere now!" And as another informant, eager to become a PR because he wanted to go back to India to join his friend's business, had explained: "I still have the job here now. I should apply now. Then I can come and go anytime." Chandary Shekhar, a thirty-three-year-old Telugu IT professional, went to the United States from India on an H-1B visa in 1997; realizing the difficulty of converting this to green card, he quickly moved on to Australia, where he became a PR and set up a company, TecSole. Soon afterward, he went to the United States, again on an H-1B visa, to work in a large IT company in order to explore the market for his own business. Navin, who was born in Maharashtra but raised in different places in India following his father's army postings, was assigned by Tata Consultancy Services to work in Australia in 1996 for one year. He applied successfully for permanent residency in Australia after he returned to India, but proceeded to the United States on an H-1B visa. After being in the United States, however, he felt that he still preferred Australia, so after returning to India in 1998 and getting married, he came back to Sydney with his wife.

Apart from being a stepping stone, Australia was also considered a base of security—to sustain IT workers' high level of mobility—due to its comparatively less volatile market, and generous welfare policies. The market slowdown, and the dot-com crash soon after, had shown Australia to be a particularly popular alternative destination to the unpredictable market in the United States. Over two days in May 2001 when I was staying with Uday, he received four desperate calls from a friend in Boston and another in New York asking how to migrate to Australia. At the same time, the Gujarati neighbor was busy collecting immigration information for his former classmate in Ahmedabad, who had studied in Sydney, moved to the United States after graduation, then lost his job and returned to India, and was planning to come to Australia again. Similarly, I came across two Indian technopreneurs who moved from the United States to Australia, not because they saw it as a place to make quick fortunes, but because its more stable market and close connections with Asia made it more likely to be a supportive environment for small IT enterprises at that time. Canada, similarly positioned in the global economy and in terms of immigration and welfare policies, was another major stepping-stone destination and base of security.

Although body-shop operators and IT workers often appeared to be plugged into a labor market of freewheeling global cybernauts, the world in their mind had by no means become a "stateless" and borderless playground. "Can I see your passport and visa?" was often the first question put to me after I introduced myself. My respondents did not intend to double check my background, but to have a look at a PRC passport and to scrutinize the British visa. When the Indian IT workers talked about their migration experience, the exact dates of getting their passports and visas were often mentioned as milestones. This somehow reminded me of M. N. Srinivas's (1967b, 54–55) description of the printers' veneration of machinery, or the tools of one's trade, in India during the Dasara festival period.[1] My informants did not feel anything remotely similar toward computers, but passports and visas—the archetypical symbols of national sovereignty that marked their mobility—were imbued with a mysterious power (and I could quite easily imagine them set amid vermilion, incense, and flowers). Australia's double role as a stepping stone and a security base further demonstrates that IT workers' strategies for multiple migrations were not only held to state policies such as immigration and employment regulations, but were also shaped by the coexisting differentials between countries that made transnational mobility profitable—a world system of body shopping.

The United States of America: "Mecca for IT People"

Despite successes brought about by multiple mobility and being fairly complacent with his life in Australia, Uday remained haunted by a dream: "I feel something in the heart every time when I hear 'United States' or 'America.'" It was sometimes so palpable that even a glimpse of the Stars and Stripes or Statue of Liberty on a television program would find him shouting like a kid: "United States of America!" As all my informants chorused, "U.S.A. equals Mecca for IT people." Senthil told me that he was just "passing through" Australia toward the United States, his single destiny, although he had been in Australia for more than two years. Sharma, the Telugu who introduced his brother-in-law to Samy, personally knew twenty families who had moved to the United States between 1998 and 2000. Another informant had made the observation that for every one hundred Indian IT arrivals in Australia, about thirty departed, almost all for the United States. The American dream of course had its firm material grounds: salaries for IT jobs in the United States were almost double Australia's, with raises every six months compared to every year, sometimes two, in Australia. But my informants also put special emphasis on the currency conversion rate. I was puzzled as to why the exchange rate would affect their everyday life, until Rajalaxmi, an IT person like her husband, spelled out her experience: "When I just came from India to Australia, it hurt. Twenty-five rupees is only one dollar. When I went from Australia to London last year, it hurt again. Do you want pounds or do you want rupees?" Thus, life choices and strategies were projected continually against a transnational, rather than local, context. The devaluation of the Australian dollar beginning in 2000 made IT workers even more eager to move to the United States.[2] Chandu, a Telugu IT student at a business college[3] in Sydney, lamented that he was "losing money every day" by staying there and he could hardly wait to take off.

Indian IT workers in the United States reported better promotion opportunities than those in Australia. I was told of a few Indians who moved to the United States and were sent back to Australia as senior managers a few years later, a position they could never have reached had they stayed. This was attributed to companies in the United States being more "performance oriented" and more "professional," so that Indian IT talents were better recognized and valued. Some argued that U.S. companies grew much faster than Australian ones and therefore always had new positions for promotion. The large number of multinational corporations headquartered in the United States was another paramount consideration for more advanced career opportunities. Ramesh, the owner of WinWin Recruiter, pointed out that most Indian IT companies set up their first overseas offices in the United States: "Why do they do that? Because you can get all the top-level people and networks in America. Then you can go *down* to other countries." Some businesspeople registered their companies in the United States, though in fact operating from Australia or India, because according to them it was easier to cut partnership deals as a "U.S. company." Uday's brother Ashok had called this the high "returning value" of going to work in the United States, that is, one would be more highly valued when *returning* to India or other countries from there.[4] By comparison, "after you stay in Australia for five years, you can't go anywhere. You will be outdated [in technology]," stated Meena, originally from Tamil Nadu, and who was determined to go to the United States.

State immigration regulations also made the United States preferable to Australia. While spouses of H-1B visa holders were allowed to enter on H-4 visas at the same time, an Australia 457 visa holder had to wait for a minimum of six months to go through the bureaucratic procedure to bring in the spouse (also on 457 visa). Young IT bachelors often worked in Australia for one or two years, returned to India to get married, then moved on to the United States together with their spouses. The fact that the H-4 visa precluded work—while the Australian 457 visa did not— was not a consideration among my informants: since the salary in the United States was high, wives *did not need* to work. Besides, for quite a few Indian IT workers that I interviewed, being able to afford a non-working wife was regarded as a measure of a man's success.[5]

Finally, since the Indian community in the United States was much bigger, it was quite common for Indian IT workers in Sydney to have more friends and family relations in the United States than in Australia. Sometimes, however, one had migrated to the United States not so much pulled by connections as pushed by families somewhere else. Residing in Sydney, Shireesha had matched her younger brother-in-law (Chandary's brother) in India to a Canada-based Indian girl. After the couple settled in Canada, she urged them to move to the United States: "They have three daughters [one to four years old]. They need money [for dowries]. I told them: You have Canadian citizenship, but just stay in Canada and don't go to America? What are you doing?" Khrishna from Kerala was helping his brother move from the Middle East to the United States when I visited him in Sydney—partly so that he could then hand over his financial responsibility for the family in India to a better paid brother!

Global Gateways: Singapore, Malaysia, and the Middle East

On the world map of Indian IT workers, Singapore and Malaysia constituted strategic gateways to the global labor market, especially for those without overseas work experience or well-recognized certification and who were, therefore, unlikely to be hired straightaway in markets like the United States. Singapore and Malaysia became global gateways first of all because both had placed great emphasis on developing the IT sector and attracting foreign IT talents. In Singapore, the former prime minister Goh Chok Tong had in 2001 described competing for and recruiting foreign talent as "a matter of life and death" for Singapore.[6] Malaysia boasted the well-known Multimedia Super Corridor (MSC), the pet project of the legendary former prime minister Dr. Mahathir Mohamad that showcased the future "K-Malaysia" (knowledge-based Malaysia). Large companies, especially those granted "MSC" status, were given latitude to hire foreign, in particular IT, specialists. These efforts were well received by the international business community. Both countries, particularly Singapore, were seen as ideal locations for headquartering multinational corporations' Asian operations, thus in turn heightening the demand for IT skills. Increasing numbers of Indian IT companies, including the high-profile Hyderabad-based Satyam, had also set up branches in Singapore in order to boost their larger presence in the Asian market.[7] For Indian hopefuls, Singapore's international position enabled them to establish contacts with multinationals, which together with the work culture emphasizing discipline and efficiency, and the English-speaking and Westernized environment, made Singapore highly valued as a résumé-enhancing destination.

Singapore and Malaysia were also suitable gateways for trial outmigrations, being only a three- to four-hour flight from southern India. Indeed, Singapore's geographical proximity to major labor-source countries such as China, the Philippines, and Indonesia has encouraged a large presence of international recruitment agents. Uday estimated that at any time between 1998 and 2000 there were hundreds of Indian-run body shops in Singapore managing thousands of IT workers. Placing Indian workers from both India and Malaysia to Singapore, and from Singapore on to Australia, Canada, and the United States was a standard business for body shops there.

Indian IT workers' main preoccupation in these gateway countries was having to look for job opportunities while at the same time coping with financial difficulties. More so than in other countries, "Indian networks" among IT workers, and linking workers and body shops, played an important role. A large proportion of Indian IT workers traveling through body shops to Singapore entered the country on social-visit visas valid for one month, in which time they combed through the large office complexes where IT placement agents, particularly Indian-run consultancies, were concentrated, hoping to secure a job. When their visas expired, some stepped into Thailand, Indonesia, and in particular Malaysia to renew their visas and re-enter Singapore to continue job hunting.[8] In the small city-state, Indian jobseekers quickly came to know and befriend one another and, while a little secretive in their pursuit of Singapore-based jobs, willingly shared information on the overseas opportunities that they tried to access from Singapore. Uday applied for jobs in eleven countries, including South Africa, Germany, and Japan, when he was working there. To save money, four or five friends might gather around just one computer at an Internet café to prepare and send off their résumés together. At other times, friends with more interview experience helped out as surrogates in international phone interviews. For these reasons, Uday told me, he had experienced Singapore as a place "for networking and bridging."

In Malaysia, many IT workers found it difficult to find jobs after arriving through body shopping because the counterpart (local) agents were not quite in the loop. When I was in Kuala Lumpur in June 2001, nearly half of an estimated three hundred Telugu IT workers there were jobless and had to rely on the "Indian network" for day-to-day survival. The influx of body-shopped arrivals had led to a peculiar Telugu enclave forming in the Brickfields area in central Kuala Lumpur: Palm Court, a four-block residential complex accommodating about 1,000 Indians, of whom Telugus were the largest regional group. It became almost obligatory for the single men at Palm Court to take in jobless new arrivals from India; in fact, body shops in Hyderabad even advised the workers they sponsored to visit Palm Court on their arrival![9] This unique ethnic clustering made it easier for those without international experience to move from India to Kuala Lumpur, but created a heavy burden on those who were already there. Like Rejeshekhar, the young Telugu IT worker I contacted through the Web site of a Telugu association in Kuala Lumpur, who asserted that no one would want to stay for more than two years: "In Malaysia you can't save. Many people ask for help. At least half [of wages] goes to . . . friends." There was, hence, considerable motivation to actively search for opportunities to move from Malaysia to other countries. In effect, both the costs and the facilitating role of these Indian networks reinforced Malaysia's position within the world system of body shopping as a popular first destination, but only for the short term.

Similar to Singapore and Malaysia, some countries in the Middle East, particularly the United Arab Emirates, Saudi Arabia, and Kuwait, were also used as gateways to the global IT labor market. These countries were relatively easy to enter, and work experience in them was thought to be helpful for getting visas to the West. And although the salaries were lower than in Western countries—a foreign programmer with two years' experience was paid about INR 40,000 a month in Kuwait, INR 30,000 in the United Arab Emirates, and INR 35,000 in Saudi Arabia in 2001[10]—they were much higher than in India. In sharp contrast to their counterparts in other countries who kept changing jobs, IT workers in the Middle East usually stayed with the same firm for four or five years because mobility was tightly regulated in the region.[11] This situation created a special business niche for agents placing IT people out of the Middle East. Asif Ali, an electrical engineer originally from Hyderabad, had worked in Saudi Arabia for three years then gone to Canada, where he became a PR. He returned to Riyadh to set up his recruitment firm, Telecomplus, with branch offices in Hyderabad, the United Kingdom, and Canada. He sent IT people from Hyderabad to Saudi Arabia, and more importantly, from Saudi Arabia to the United Kingdom and Canada.

South Africa was another global gateway, notably having eased its immigration laws in the late 1990s in order to attract skilled foreigners from countries such as India and Russia after losing over 1 million skilled personnel in the aftermath of the dissolution of the apartheid system in the 1990s. Body shops in Hyderabad typically sent IT workers to South Africa as project teams of four or five workers. Workers went to South Africa not only for jobs, but also to move on to the "white" commonwealth countries, which was much easier done from South Africa than from India. Some first moved around to countries close to South Africa, particularly to Kenya and Zimbabwe where a number of Indian companies had long-standing connections, then moved to South Africa, hoping to eventually go to Australia or Canada.

U.S. Satellites: The Caribbean and Latin America

From the end of the 1990s, a sizeable group of Indian IT professionals were body shopped to the Caribbean countries, particularly Jamaica. This was triggered by the so-called nearshore business model of some U.S. IT companies, in which they exported tasks to nearby locations where, at the same time, they imported labor from elsewhere. Jamaica has ideal attributes for this business model—cheap land, low wages, and a geographical proximity that allows U.S.-based managers to visit frequently.[12] The nearshore model was anticipated in 2000 by the Jamaican government to create as many as 40,000 IT jobs, though mostly in the low-end sector such as call centers.[13] Body shops, quick to spot this trend, entered these new regional markets, placing workers in software-development houses as well as in IT-enabled services (such as call centers) for maintenance tasks. Most placements to Jamaica were for middle-level jobs—among the senior-level American managers who commuted between Jamaica and the United States, and junior level staff mainly comprising local Jamaicans. Indian middle-level workers lived in gated compounds rented by the company and had hardly any connection with the local society.

Some Indian IT firms also set up joint software-development projects with local Jamaican partners, targeting the government as main customers, or, like Tinager, a Chennai-based IT firm, set up a medical-transcription firm there to serve U.S. hospitals and hired fifteen workers from India in late 2000. These Indian companies were also engaged in placing IT workers from India in Jamaica, and of course, from Jamaica to the United States. According to my informants, to say "I am going to Jamaica" meant little; instead, they positioned themselves as working for U.S. companies that happened to be located in Jamaica. Nevertheless, Jamaica remained a popular choice, especially for single workers, due to its proximity to the United States and a notion that migrating from Jamaica to the United Kingdom was particularly smooth sailing.

New Frontiers: "Sind Sie Inder?"[14] and "Is There a German Dream?"

The new frontiers are those that only recently opened their doors to foreign IT workers. In Europe, these were countries such as Germany, France, Italy, Sweden, Austria, and Switzerland, which are in fact part of the "old world" of modern European outmigration; and in Asia they include Japan, South Korea, Taiwan, and Hong Kong. Germany, a former "zero immigration" country with a ban on overseas recruitment since 1973, provides a proverbial example of this opening up. In May 2000, and coming at a time when the unemployment rate was as high as 9.6 percent, the government approved a new program to allow up to 20,000 non-European Union IT workers to be issued work visas valid for two years and renewable for up to five years.[15] What was greeted as a "paradigm change"[16] by Rita Süssmuth, chairperson of the newly established German independent commission on immigration, had anti-immigration groups in Germany shouting "*kinder statt Inder*" (children not Indians)[17] on the one hand, and IT firms in the United States crying out for the government to "get the Indians *here* before Germany grabs them," on the other.[18] Disappointing for the German market, however, a year after August 1, 2000, when the program took effect, only 8,556 green cards had been issued, significantly lower than expected, with Indians accounting for about 20 percent (1,782), exceeding the numbers from the states of the erstwhile Soviet Union (1,198) and Romania (736).[19] More than 60 percent of the visas were sponsored by companies with fewer than one hundred employees,[20] which may well have included body shops or agents alike.

In Asia, Japan in February 2001 decided to offer multiple-entry short-term visas specifically for Indian IT professionals, allowing stays of up to ninety days and valid for three years (Ministry of Foreign Affairs, Japan 2001). This was a breakthrough because until then Japanese companies did not recognize any foreign computer-engineering qualifications.[21] Japan also planned to train 1,000 Indian IT professionals in the Japanese language and business practices over 2000–2003 (Ministry of Foreign Affairs, Japan 2001).When I was in Hyderabad in 2001, South Korea was a starred new frontier where IT workers were reported to be much in demand—over 20,000 were needed in the year 2001–2002.[22] Huge banner advertisements for jobs in Korea, promising as much as USD 3,000 a month, hung from the doors of some body shops.

Moving to these new frontier countries did not necessarily mean working in Asian or European companies, since many job vacancies were in U.S. subsidiaries in these countries. Nor did it mean working in a non-English-speaking Asian or European environment. For example, many of the demands for IT labor from Korean IT companies were from their export divisions where English was the working language. In this sense, Germany's strong but domestic IT industry might have been its weakness in attracting workers from overseas, because a job in a local German company was less résumé-enhancing than working for a U.S. firm in South Korea. Some authors attributed the relatively small numbers of Indian IT professionals moving to Germany to the lack of Indian networks there (e.g., Poros 2001). But it is more likely that the Indians *chose not to create* German networks. As Murali Krishna Devarakonda, director of Immigrants Support Network (basically an Indian IT H-1B holders' association based in Santa Clara, California, to lobby for a more liberal green-card system for H-1Bs), asked: "Is there a German dream"?[23] No matter where they went, most IT Indians moved to non-English-speaking industrialized countries with a clear plan to move out soon.[24]

Thus, the emergence of the world system of body shopping was in part a process whereby IT workers compared the situations in different countries carefully and pursued their interests by capitalizing on the differences across countries. In this system, Singapore and Malaysia prepared junior Indian IT workers for entering the higher sectors of the global market. IT workers jumped into the high-paying but volatile job market in the United States from Australia—after developing their skills by utilizing the stable employment relations there and earning permanent residency, which gave them a base to return to. The flexible employment relations and "risk-taking culture" of industry leaders in the United States had been sustained only because part of the cost for such uncertainties (resulting in, for instance, the lack of corporate investment in human capital) was exported to secure bases such as Australia. Thus, in the world business of body shopping, the dynamics between the United States, holding zones like Australia, and the base of production of the primary material (IT labor force), India, correspond well to those of core, semi-periphery, and periphery in the world system defined by Immanuel Wallerstein and his associates.

Ending Remarks
The "Indian Triangle" in the Global IT Industry

Information technology is widely regarded as a major cause of a series of fundamental social changes that we are going through, and is even expected to usher in an entirely new epoch in human history. Indeed, "IT revolution" is the only revolution that we can talk about today. This technology euphoria, however, often leaves unexamined the social processes through which the technologies are generated and implemented. This book has demonstrated that IT is powerful precisely because it is integrated with, therefore facilitating of, other operations, and the technologies would not have been developed without the demand for new paradigms of production and management, the changes in our work pattern and even life style, and, of no less importance, the ideology of the New Economy. IT is thus itself a social construct.

The IT industry as a social process is in a sense new, as evidenced by its global scope, its flexibility, and most notably its volatility. But lying beneath the novelty, this book argues, is nothing mysterious: it is a special pattern of labor management, including the production, mobilization, and control of labor. Labor is crucial here not only because the IT industry is labor intensive (which is itself a deliberate rather than "natural" choice), but also because creating a labor system responsive to extremely volatile capital movement on a global scale is something hardly ever before experienced by other industries. Body shopping thus provides a strategic case for scrutinizing how the increasingly "abstract" economy is constructed through reworking concrete human relations and institutional arrangements.

In describing the internal workings of the body-shopping practice, this book has used three terms: "ethnicization," "individualization," and "transnationalization." The concept ethnicization indicates that ethnic (Indian) networks formed the backbone of the business, particularly in mobilizing a potential labor force, maintaining labor relations internal to body shops, and expanding body shops' global networks. Yet the body-shopping business was not based on preexisting, closely knit ethnic networks. Kinship or old-town connections, for instance, hardly played any role, whereas "professionalism" was the key word. It was workers' individualist attitudes, meritocratic ideologies, and ambitions to move up that made body shopping sustainable, thus the word "individualization." The term "transnationalization" points to the institutional significance of body shopping: the practice came into being because it enabled capital to utilize and manage labor globally; and, to fulfil this function, body shops operated transnationally. This explains why ethnicization and individualization can work together. The body-shopping practice was "ethnicized" not because of "Indian culture," but because India became a production center of labor force for the world IT industry. For this reason, individualistic calculation rather than bounded solidarity or Indian cultural attributes underlined body shops' apparently ethnicized operation. Body shopping turned IT workers into "free" atoms in the market as envisaged by neoclassical economics; yet precisely in order to achieve this, the business had to base itself on ethnic networks for its operation and on the socioeconomic structure of local India for resources. Body shopping helped free the global IT industry from state regulation on migration (though by no means completely), but it was itself strongly conditioned by the established international economic order.

What does body shopping mean for India? What is possible with awareness of all the stories? These final remarks provide a normative critique of the subject by clarifying how body shopping is related to the Indian "IT miracle" and to the larger society of India.

The spectacular growth of India's IT industry has attracted much attention worldwide, and Atal Bihari Vajpayee (1998), the former Indian prime minister, claimed that "*India's Tomorrow is IT.*" The simultaneous and mutually facilitating development of the two established Indian IT sectors—the one within India as represented by companies such as Infosys, Wipro, and Satyam, and the one run by Indians overseas[1]—has been widely cited as a celebratory symbol of globalization. Body shopping, however, is left unmentioned or is merely touched upon as a transitory phenomenon soon to disappear. Moving away from this mainstream narrative, my study suggests that the two established Indian IT sectors in fact relied on the "informal IT sector" comprising body shops, training institutes, and unemployed or semi-employed IT workers in India. The formal IT sector in India, the informal sector, and the overseas sector were related through the mobility of labor, and they thus constituted an "Indian triangle" in the global IT industry (see figure 2). In this triangle, the low-tier informal sector and the high tier in India were not linked directly, but were connected through a transnational circuit: body shops accumulated resources by sending labor overseas and then moved up into the high tier in India. Acting as the key agent of organizing the locally produced labor and delivering it to the global market, the low-tier informal sector effectively siphoned off local resources—embodied in the labor force—to sustain the entire triangle. Without the informal sector, the overseas sector would immediately lose its competitive edge (i.e., skilled and cheap labor), and the high-tier sector in India would be deprived not only of labor, but also of the finance capital brought in by the small players when they moved up. Precisely because the established IT sectors were based on this informal sector that in turn relied on the local society—a large resource base, the Indian IT industry could develop rapidly and survive well the global market slowdown at the beginning of the new millennium. The informal sector transferred business risks to workers and even extracted financial contributions from them by selling jobs.

The Indian triangle constitutes a sharp contrast with the East Asian experience of local-global economic relations. At first glance, the Indian IT industry is similar to the export-oriented manufacturing in East Asia that had been the basis for the economic miracle of the region, and indeed, the IT industry is often predicted to have even stronger spin-over effects given its higher added value. But in East Asia, local-global connection tends to be achieved through vertical cross-firm links in which large firms obtain orders from the global market, then subcontract some of the tasks to smaller companies. Small industries and even households can thus join the global market, which is particularly obvious in Taiwan and parts of mainland China where rural industries have been well developed. Exploitive as it may be, this connection generated employment and business opportunities on a wide basis, and contributed to the rapid and largely equitable development with a very high level of employment (Fields 1984; Barrett and Chin 1987). In India's IT triangle, however, little wealth is passed downward from the global to the local, though conversely, value is pumped both upward and outward (to the West) from the local. The triangle also indicates that the IT industry is highly dependent on the global market with very limited linkage to the domestic economy.

Besides the mechanisms of labor production discussed in the book such as private-sector education and the institution of dowry, unequal gender and class relations were also crucial. In Andhra Pradesh, only 1 percent of women were employed in the organized sector in 1998 (DES, 413, table 28.17), and most women's work, essential for the production and repro-

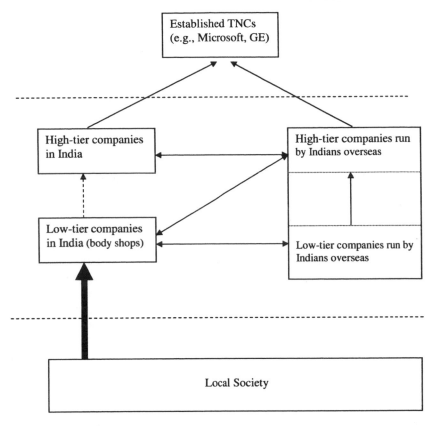

FIGURE 2. The "Indian Triangle" in the Global IT Industry

duction of labor, is not calculated in monetary terms. Servants, mostly women, are common.[2] Almost all the medium- and small-sized companies in India that I visited, both body shops and specialist IT firms, hired boys between eight and fifteen years old to serve tea, buy lunch, and mop the floor, with a monthly wage of between five hundred and eight hundred rupees.[3] It is the invisible and undervalued work of these women, children, and many other men that enables the Indian IT labor force to be produced at very low costs. It is therefore hard to assess who—an "untouchable" woman cleaner, a Kamma H-1B man, or an American venture capitalist—contributed the most to Silicon Valley's glory. Ravichandary, son of a politician in India working as a telephone receptionist at Advance Technology in Sydney, welcomed the news that HIV was becoming a major epidemic in India: "Let them die! If half of the population die, India will be much better." Although few put it so blatantly, most Indian IT people whom I met agreed that India could be as prosperous as the United

113

States if it halved its population and doubled the number of those as "productive" and "competitive" as themselves. However, if the half population they saw as a burden were eliminated, the IT people would be wiped out of the global market as well. It is not their English proficiency or the developed tertiary education that made Indian IT people competitive, as the mainstream media and most of my respondents believed; rather, it is the severe class, caste, and gender inequality that allows surplus value to be especially efficiently mobilized and concentrated on a small elite group to produce competitive IT labor.[4]

In assessing the social consequences of IT and globalization, existing academic thinking and practical agendas of international agencies and national governments have been dominated by the concern about "digital divide" or more broadly "social exclusion." It is feared that the new technologies and the trend of disembedding would bring about a polarization in which the elite become global and the poor stuck in the local (e.g., Bauman 1998; Hoogvelt 1997). Castells (2001, 277) made this point particularly sharply by asserting that the Internet would "divide people around the planet, but no longer along the North/South cleavage, but between those connected in the global networks of value-making, around nodes unevenly dotting the world, and those switched off from these networks." The global production system would thus become "composed simultaneously of highly valuable and productive people and places, and by those who are not so, or not any longer, while still being there" (Castells 2001, 266). The body-shopping case demonstrates, however, that how insiders and outsiders of high-tech reach are *connected* is more significant than how they are divided. No human being has become nonvaluable or nonproductive, and the question is precisely why some appear so. A large population has been *deemed* irrelevant in globalization despite its direct and indirect contributions to global wealth. The ideological, discursive, and often political exclusion was critical for maintaining the Indian triangle and particularly its relationship to local society. It is this ideology that mystifies the power of technology, glorifies and naturalizes the suddenly acquired wealth of the IT people, and justifies the large amount of resources allocated to the IT sector. It is exactly through ideological exclusion that unequal material connections are reinforced.

But this book should not be read as an antiglobalization argument calling for a crackdown on body shops. Stricter regulations in the destination country may intensify the friction between the state and the corporate demand for flexible labor, only making body shopping as a means of removing the friction more indispensable. Tightening up may also push body shops to become more abusive: it is commonly known that body shops in the United States were "tougher" with workers than those in Australia mainly because of the more restrictive rules. For India, although body shopping intensified existing inequities, IT did bring new developmental dynamics to some societal sectors. IT professionals have evidently become a major social force for economic changes and a significant political constituency. A key purpose of this book is precisely to urge them to recognize the social basis of their success and to redefine their relationship to the local society. Much more should be done than setting up Internet cafés in villages as charitable work!

Appendix Essay
The Remembered Fieldwork Sites:
Impressions and Images

Since this book is constructed to delineate the dynamisms of a "system" spanning multiple countries, it does not contain much material to bring to the reader a flavor of the research sites and a sense of "being there" as most ethnographies do. For the same reason, my informants' experiences are fragmented in different chapters to serve my thematically organized argument. To compensate for this, I attempt in the appendix essay to provide a thumbnail sketch of the typical settings—as I felt and remember them—in which I interacted with my respondents. The biographical index following this describes the life stories of my Indian informants; most of this information is not included in the chapters. For those with whom I am still in touch, the information is updated to mid-2005.

Sydney

Following what I learned from numerous anthropological studies on immigrants, I identified the Indian grocery store as a strategic "node" of the "diasporic space"—my initial research focus. Tambi, a Tamil in his forties running a large grocery in Ashfield, was thus fortunate enough to become one of my first informants. Tambi urged me to learn Tamil and asked why I had not studied IT, which, he said, had now lifted the Indian civilization

FIGURE 3. By Indians and for Indians: An IT Training Class Provided by a Body Shop in Sydney (December 2001)

to a new height. Tambi had another good reason to think highly of IT. His grocery neighbored the body shop Advance Technology Institute; students of IT training courses came down to his shop to buy snacks during breaks, and workers shopped there on their way home in the evening. In May 2001, when Tambi's wife was ill and the family could not come to the shop every day, he hired two part-time shop assistants—both benched IT professionals from Advance Technology.

I did not study either Tamil or IT, but I visited IT training classes and weekend Indian language schools frequently. The latter, originally identified as another strategic node of the diasporic space, turned out to be an important venue for establishing contacts with body-shop operators. Running weekend Telugu or Tamil schools for children was a key activity of the associations dominated by body-shop operators and other middle-aged immigrants, and chatting while waiting for the kids was the main means through which IT professionals caught up with one another. Body-shop operators who were active members in the associations took charge

FIGURE 4. A Street in Ashfield (February 2001)

of the running of the class in turn, brought in tea, coffee, and biscuits to share. Their job here was, as they often said, "to serve the community."

Sydney does not have clear concentrations of Telugu IT professionals, body shops, or even Indian populations in general. I did my fieldwork mainly in three dispersed areas: the central business districts in downtown and north Sydney where I lunched with IT workers or waited for them after work; the central west part where I lived with Uday and others; and the far west of Sydney. Ashfield in the central west was a favorite place for young Indian IT workers to hang around due to its convenient transport, Asian-style food courts, and, good for Tambi, Indian groceries; and many recent arrivals on 457 visas jointly rented flats in the area. Strathfield, a posh suburb in the central west, was called *Strathpura* ("*pura*" means "town" in Hindi) in the 1970s and 1980s in the Indian community because of the large number of Indian residents, particularly physicians, there. By the time of my fieldwork, about one hundred Indian families had houses in Strathfield proper, much fewer than in other suburbs. The periphery of Strathfield close to the railway station where Uday and I lived looks completely different, and the house rent was much lower. Uday never visited any Indian family in *Strathpura* and always asked me what the houses looked like after I returned from interviews. Finally, many body-shop operators bought or built houses for their own use, or rented

119

flats to sublet to workers in far west Sydney (including the suburbs of Liverpool, Blacktown, and Parramatta) where real estate price is low. Quite a few body-shop offices were also located there due to the low rent.

Kuala Lumpur

I had one of the most striking encounters during my fieldwork in the Kuala Lumpur airport on my way from Sydney to Hyderabad. A young Telugu IT worker, being told that transiting passengers could stay in Malaysia without visas for seventy-two hours, had flown in with a return ticket from Melbourne, hoping to zoom around and get a job in Kuala Lumpur within this time. It was only at the immigration checkpoint, where we were both queuing, that he found out that the seventy-two-hour-stay only applied to passengers heading for a third country; he could not even leave the airport and had to take the very next flight back to Melbourne, where he had been benched for the previous four months. With all seriousness, he asked me whether I could "get a break" for him in China.

This unlucky Telugu was one of the many Indian IT workers coming to Malaysia to look for jobs, mostly on business or social-visit visas; and there were so many such that they had become a special target of local police for demanding bribes in Kuala Lumpur. A friend of Rejeshekhar (whom I contacted through the Internet from Sydney) told us of how he was stopped by a policeman on the street who asked for his passport and work-permit card; when he said he did not have the documents with him, the policeman asked: "Are you here on business visa?" He said yes, and the policeman shouted immediately: "You are working here, right?!" The friend insisted that he was not (the business visa does not allow the holder to be employed in Malaysia), and the policeman asked him to "walk with" him—implying that he was in custody. The friend stopped the "walk" with a payment of MYR 600 after some bargaining. He cried for two days—he had worked for only less than two months and was now deprived of all his savings.

While IT workers from India were busy hunting for jobs, the Malaysian Indian Congress (MIC), the third biggest national political party, was also preoccupied with IT, but for a very different reason. Since no election was scheduled in 2001, the party identified Tamil Internet 2001—an international conference to promote Tamil Internet and Tamil computing—as a priority for that year. In conjunction with the conference, MIC planned a nationwide Net for Life campaign to help the Indian community in Malaysia equipped with IT, and an international exhibition DuniaWeb 2001 to showcase Internet technologies, products, and services from Malaysia, Singapore, the Indian subcontinent, and other regions. One of

FIGURE 5. A Bedroom Shared by Four Telugu IT Workers in Palm Court (June 2001)

the key sponsors of Tamil Internet 2001, a large IT company with the Multimedia Super Corridor status, had over 80 per cent of its workforce from India (interviews with three MIC cadres, June 2 and 3, 2001).

Hyderabad

Upon arrival in Hyderabad, I was immediately advised to visit Aditya Enclave on Ameerpet Road, a residential complex built in 1993 but basically occupied by small IT consultancies by 1998. When I was taking photos of the numerous banners of advertisements for software programs, training courses, and job placements covering the building, I was stopped by a security guard in uniform and escorted to a telephone shop to see the owner—who was also the president of the Enclave Residents Association. Intervention by a security guard of this kind is very rare in Hyderabad. It transpired that the IT consultancies in the complex had set up the association to supervise cleaning and security staff—concern about security seemed to be an indispensable part of the portfolio of IT companies worldwide even before September 11, 2001. After a brief investigation of

121

FIGURE 6. The Façade of Aditya Enclave on Ameerpet Road, Hyderabad (July 2001)

me, the president told his story. The president, an electrician, bought a shop space on the ground floor in Block B of the Enclave in 1993 to run his electronic contract business. In 1996, with the advent of IT consultancies and IT people to the Enclave, he converted the shop to a telephone booth cum Internet café. At the same time, he bought another shop across the road (outside of the Enclave) to provide services of photocopying, graphic design, and printing, almost exclusively catering to the needs of IT consultancies and training institutes to inundate passengers and students with brochures and handouts. The Enclave has thirty-two shop spaces in each block, and many of them had changed from clinics, pharmacies, groceries, and clothing stalls to businesses related to IT or emigration, such as photo shops for passport shots and bookshops selling maps.

The colonialization of the Enclave by IT consultancies ballooned the resale price of a flat from INR 8 lakhs in 1993 to INR 14–15 lakhs by the late 1990s, and the monthly rent from INR 4–5,000 for 1,000 square feet to INR 8–10,000 over the same period. The global market slowdown— about five IT companies in Block B closed down in the first half of 2001— however, did not bring down the price at all because the slowdown also forced many companies to move from more expensive areas, such as the Begampet district, to the Enclave. More Internet cafés were also set up

FIGURE 7. Cyber Tower, the Main Building of HITEC City (August 2001)

catering to the needs of unemployed IT people when some IT training courses and consultancies died out. Encouraged by the price hike, the developer of the Enclave invested in 2001 in two more residential complexes in the same area targeting NRIs. The inflation of the real-estate price in Hyderabad was however only moderate compared to that of Bangalore: a few non-IT residents from Bangalore complained to me that the

123

FIGURE 8. A Typical House in Jubilee Hills (August 2001)

city had become so expensive because of the IT industry that they had to either move into IT or move out of the city!

Aditya Enclave has its main entrances at Satyam Camine Talaies Road off Ameerpet Road, a narrow street about eighty feet long hosting twelve big training institutes in 2001—one of them had 450 students every day. The large crowd brought about vibrant businesses. A couple of student-recruitment agents for overseas universities (particularly for those in Australia and New Zealand), for example, moved in from elsewhere, as one agent owner explained: the unemployed IT "floating people" in the area provided ideal potential clients. Indeed, many long-term potential emigrants went abroad as students after failing to find jobs both in India and overseas. I also ran into a neatly dressed young man busy with market research for his business plan of delivering services to the families left behind in Hyderabad by IT NRIs for a charge of USD 25 a month—a price decided "scientifically" based on his interviews with potential IT NRIs. He had no doubt about the prospect of the plan—by buying his services, he reckoned, the overseas IT son could show both social status and filial piety to the extended family—and the idea would be worth millions of dollars if he had access to VCs (venture capitalists) in the United States. Ameerpet was not all about business and IT though; there were also corners of romance. I was asked by a couple of male students to approach female classmates, and a female teacher on one occasion, to pull strings on their behalf.

FIGURE 9. Visu Consultancy in Bhimavaram (August 2001)

The "real" IT place in Hyderabad, however, as most of my informants contended, is the HITEC City in Madhapur district. Standing for Hyderabad Information Technology and Engineering Consultancy City, HITEC City is the largest IT industry park in India, offering "world class state-of-the-art IT infrastructure," as described in numerous official and corporate documents. A showcase of "Cyberabad," HITEC City was even touted as a tourist site in official brochures of Hyderabad and Andhra Pradesh. Adjunct to HITEC City is Jubilee Hills, a suburb chock full of Southern California–style mansions, many belonging to successful IT people.

Rural Andhra

My field research itinerary almost exactly reversed the typical journey of a Telugu IT professional: starting from Sydney, transiting in Kuala Lumpur, and then Hyderabad, before the final leg in villages in coastal Andhra. The sense of "reversal" came to me particularly strongly when I passed auto-rickshaws ferrying two or three schoolchildren, in most cases boys, on my way to villages. They were usually from upper-middle-class families and were commuting to their private schools in town from home villages. While education at government village schools (panchayat union elementary school) is free, private schools in town cost approximately

125

FIGURE 10. *Kachcha* Houses (Mud Huts) in a So-called Low-Caste Area in a West Godavari Village (August 2001)

INR 500 per month, and private intermediate colleges (ages eleven to twelve) INR 25,000 per year (including expenses). The singular attraction of private schools is that the courses are taught in English—instruction in government schools is in Telugu. Coastal Andhra, as one informant put it, is "education rich." Bhimavaram, a town with a population just over 140,000, had about twenty private schools. Some villages, such as Peuugonda (about twenty-two kilometers away from Tanuku town), have their own colleges. According to a lecturer at the locally prestigious West Godavari Bhimavaram College, the education quality of colleges in the coastal districts was no lower than in Hyderabad, and people moved to the capital mainly for the lifestyle and to go abroad.

But the chances of going abroad directly from coastal districts were in fact higher than people usually thought, as pointed out by Vasan, the twenty-six-year-old owner and operator of the Bhimavaram franchise of Visu Consultancy, a large training institute cum IT consultancy headquartered in Hyderabad. Vasan's key business was coaching for Toefl (the Test of English as a Foreign Language) and GRE examinations, but the consultancy also provided IT training and had recently started body shopping by linking with agents in Hyderabad. Quite different from what I assumed, often it was Vasan who provided the leads for overseas jobs—capitalizing on his connections with a large number of IT professionals

FIGURE 11. A Corner of a Village (August 2001)

overseas originated from Bhimavaram—and his Hyderabad associates handled the paperwork. In two cases his Hyderabad associates even placed workers in the United States through his connections!

How Vasan had set up the franchise provided another example of how my journey reversed that of the IT people's. Vasan's parents in a village near Bhimavaram had been farming a 54-acre prawn tank, (a man-made pond [often raised from the ground] for prawns), including 4 acres of their own and 50 leased at a rent of INR 20,000 per acre per year. They had to borrow money from private lenders at the interest rate of 24 percent per year—a common practice for farmers in this region since they did not have enough security to take bank loans, and big landowners hardly invested in agribusiness. Partly due to the lack of capital, the family, like other farmers, could not sustain the quality of prawns (which were mainly exported to the United States and Japan), and consequently the price dropped from INR 460 per kilogram in 2000 to INR 240 in June 2001. Losing confidence in agribusiness, the family decided to follow the global trend to IT, and invested INR 5 lakhs to set up the consultancy.

In Tanuku and Bhimavaram I made a few friends quickly, mainly students of local colleges, and I followed them to their home villages. Mahathi, one of them and a second year undergraduate student in English, looked reluctant when I suggested a cup of tea when passing a small tea house at the junction of his home village and the main road. His reluctance made sense only when I learned, later, that Mahathi and his fellow

127

caste members, of the so-called low categories, were not supposed to enter the tea house and could drink outside only. The village is divided into three or four streets with completely different houses and facilities and with clearcut boundaries between: good streets for the so-called high caste, and bad for the so-called low ones.

In a marketplace in Tanuku, I ran into a former member of the Communist Party of India (Marxist). After learning that I was from China, he ordered his nephew, a shopkeeper, to sell me the cotton kurta that I liked at the price he'd purchased it at. He also gave me the contact information for one of his comrades, a retired school principal now residing in a village close to Tanuku. After a long talk about political philosophy, world affairs, and the complex relations within the communist movements in India, the former principal dispatched his granddaughter, Rani, a master's student in chemistry, to show me the village. Rani took up the assignment happily but when we walked into a different street, her steps became stiff, her face red, and her head bowed: the street was suddenly full of widened eyes. Rani told me that as a Kamma woman (Kammas had been the dominant group in the Communist Party of India in Andhra Pradesh, particularly in the later Communist Party of India (Marxist), and communism in Andhra was thus once dubbed Kammanism), she was not supposed to walk in low-caste streets since it would pollute her. Although for brave young people like her, walking through was no longer an absolute taboo, changing residence across streets is still unthinkable. Low-caste families with professionals or well-off emigrants could not move to high-caste streets with better infrastructures, and thus migrating to towns and cities became a more feasible option.

Biographical Index of Informants

Aberami, a Tamil in his sixties, graduated from Imperial College in London in 1977. He turned down a contract job offer in the United Kingdom because he was "super-confident that I will find a better job" with the British degree. On the flight from London to New Delhi he saw in a magazine a job advertisement from a Melbourne-based university and thus went to Australia in 1978. I got his contact details from Chaya, but secured the first interview through his university colleague, a sociologist. (65–67)

Arun, a Telugu 457 visa holder and Venu's roommate, went to Australia in 1999 sponsored by Advance Technology. He then went to Germany from Australia through a body shop in India in 2001; from there he submitted his application for Australian permanent residency and went back to Australia and became a PR in 2002. (73, 96)

Ashok, Uday's brother, born in 1975, graduated from Osmania University in Hyderabad in 1997, and subsequently joined Baan, a Dutch software company with a large branch in Hyderabad. In Baan he was assigned to Taiwan, the United States, and the Netherlands. He joined Uday in Sydney in March 2001 as a permanent skilled migrant. (xvi, 25, 38, 61, 93, 95, 103)

Ashwin, Uday's cousin, went to the United States on an H-1B visa in 1998. (29, 40)

Asif Ali, a Muslim Hyderabadi, worked in Saudi Arabia as a chemical engineer (1992–1995), migrated to Canada in 1995, and returned to Riyadh to set up his recruitment business in 1999. (106)

Chandary and **Shireesha,** a Telugu couple in their late forties, had worked as managers in large companies in various places in India. In 1985 they went to Singapore when Chandary was introduced by a former Tamil colleague to a company there. Chandary was posted to Malaysia where he worked until 1988, when they migrated to Australia following another Indian friend's suggestion. Learning that female applicants are normally preferred, Shireesha lodged the immigration application with Chandary's status indicated as that of dependent. In Sydney Shireesha worked in a community welfare office, and Chandary worked in a large mining company. On May 9, 1993—a date that Chandary will remember forever—he was laid off and was given as a memento a briefcase with the company's logo. The layoff was a huge shock to Chandary, so much so that all the details remain clear to him. He kept the briefcase in the upper right drawer of his desk (so that he could see it almost every day) to remind himself of this bitter experience and the hardship of being an immigrant. When I asked him why he decided to set up his own business, he took out the briefcase and waved it: "They gave me *this* when they fired me. I will never forget." He started the CSR Holding Pty., a software-development firm, partnered with a Polish colleague who was also laid off. In 1997 CSR worked on a project for the Australian branch of DMR Consulting (a Canadian multinational belonging to the Fujitsu Group), which recommended CSR to DMR Consulting United States in 1998, which in turn outsourced to CSR a project that needed forty workers. Chandary and Shireesha launched a large recruitment program in India for this and thus started their body-shopping business. Almost all my earliest informants—Indian association leaders, academics, and newspaper editors—suggested I visit Chandary because he had been very active in the Telugu community, but no one mentioned the body-shopping business to me. (55, 70, 104)

Chandary Shekhar, a thirty-three-year-old Telugu, went to the United States in 1997, then to Australia through a body-shop operator whom he ran into at a wedding party in Hyderabad, and became an Australian PR in 1998. He went to the United States in 1999. (101)

Chandu, twenty-five years old, from East Godavari, went to Australia as a business-college student in 2000 after his applications for colleges in the United States were rejected. He helped his sister go to New Zealand as a student in 2001, but he returned to India in 2002. (103)

Chawdary, a well-connected Telugu man in his forties, was a public-relations manager of a medium-sized IT company in Hyderabad set up by his remote relatives. Chawdary's grandfather was a freedom fighter and was arrested by the British in 1921, and his father was a businessman cum politician in Andhra Pradesh. Chawdary's main job at the company was to deal with two groups of people: politicians and street gangs (*gunda* in Telugu). Local politicians, known or unknown to the company, often called up asking for jobs for their clients, and only people like Chawdary, who knew how deep the water is, could handle this. Gangs came around at festival times asking for donations, and refusals could lead to violent revenge. Chawdary needed to identify quickly which politician was behind the gang and which police officer could be trusted. Chawdary was also using his own money to run a newspaper *Samaikya Andhra* (United Andhra) to counter the campaign for an independent Telengana state. The campaign, mainly led by Reddies and in a full swing during my fieldwork in 2001, justified the cause by pointing to the huge disparity between the Telengana region and the rest of Andhra Pradesh, which was much exacerbated by the IT industry concentrated in Hyderabad and coastal Andhra. (25)

Chaya, born in the 1960s in Chennai and an IIT graduate, went to Australia for the first time in 1982 for a master's course as part of the "Colombo Plan" (which was launched in 1950 with the purpose of enhancing cooperation among the commonwealth countries in Asia Pacific and preventing the spread of communism). In 1984 Chaya returned to India to work at the high-profile National Informatics Center in New Delhi headed by Dr. N. Seshagiri, a returned NRI and a friend of Mr. Rajiv Gandhi's. Unable and unwilling to speak Hindi, Chaya found it difficult to live in Delhi and went to Australia again in 1988 as a Ph.D. student. After graduation Chaya worked at Norvel, Cisco, and IBM, and in 1998 quit his job to start his company, Sysway. Due to difficulties in business, he joined Singdin, a company set up by Aberami and Rangarajan, and in 1999 he went back to work for IBM at the same time continuing the businesses of both Sysway and Singdin. Half a year later, he again left IBM, and set up a new consultancy, TrueInfo, as a joint venture with his friend Nadarajah. TrueInfo was closed down in 2002. In 2005 Chaya was working as an employee of a large IT company in Sydney and at the same time as a part-time consultant specializing in SAP under the name Sysway.

Chaya had been active in the Tamil community since he was a master's student. Dissatisfied with the Tamil association led by secular physicians at that time, he joined a few Brahmin Tamils (although non-Brahmin

131

himself) to set up a new association. While the earlier association welcomed anyone interested regardless of religious background but discouraged the participation of those from Sri Lanka, Chaya's association was tightly linked to the Sri Lanka community, which devoted most of its efforts to building the Sri Venkateswara Temple, possibly the largest Hindu temple in the state of New South Wales. Chaya's wife, from the same district in Chennai as Chaya, is a primary school teacher, and they had an eleven-year-old son in 2001. (56, 65–67, 72, 79, 87–89, 91)

David, an engineer from Kerala in his forties, a Christian, went to Australia in 1994 as a skilled immigrant. He was active in organizing Kerala functions—I met him the first time at a children's performance. He started his body shop, World Digital, in 1998, as an alternative to going to the United States: "I can't go [to the United States] because of my family. . . . Australia is nothing compared to there [the United States]. But doing a business is Okay. At least you have something challenging." World Digital was closed down in 2002, and David was working in an Australian IT company as a contract consultant in 2005. (90–92, 94, 98, 100)

Ganga, a twenty-eight-year-old Telugu doctor working in Saudi Arabia in 2000, studied IT in 2002 in India after his contract in Saudi Arabia was up, and went to New Zealand as a master's student in media studies in 2004. He said that he would rather commit suicide than stay in India because his "lifestyle could not fit in." (40)

Gangadharam Atturu, a twenty-three-year-old Raju from West Godavari, was on his way home from the United States after being laid off when I met him on the airplane from Kuala Lumpur to Hyderabad. After finishing his undergraduate course in Andhra Pradesh in 1998, one of his school friends in the United States forwarded his résumé to a few Indian body shops there, and he was asked to go to the United States within ten days of a telephone interview in February 2000. When he rushed there, however, the project was canceled. His flatmates were laid off one by one: one who had worked at IBM now worked in a sandwich bar, and another, who had worked at Citibank, found a job in a petrol station. Atturu's former supervisor in the Hyderabad company, who went to the United States two months before Atturu, was also working in a petrol station—he had to stay on because he had just bought an expensive car and was paying a monthly mortgage. Atturu returned to India in June 2001 with a return ticket—at that time it was rumored in both the United States and India that the United States market would pick up again in December 2001 with the passing of a new tax policy. Atturu stayed in Hyderabad with his sister, who married there, and he worked as a part-time instructor in a training institute. In early 2002 he learned that the body shop that had sponsored him in the United States had closed down and thus his H-1B visa was no longer valid; he then moved to Gujarat where a friend of his brother-in-law's promised to help with a job. (17, 36)

Gopal, a twenty-four-year-old Telugu, went to the same high school as Uday. He asked Uday for advice about going abroad in 2000, and Uday referred him to Ravinder, and this was how I got in touch with him. He went to Hong Kong through a body shop in 2002. (24, 46)

Govindan, a thirty-five-year-old IT professional originally from Orisa, worked as a senior programmer in Bangalore, and was then assigned to Jamaica in 1998. In 2000 when his wife became pregnant, they went to Australia with the sponsorship of David (Digital World), introduced by Govindan's former classmate. The couple settled in Melbourne in 2005. (90–93, 98)

Imitaz, an architect from Mumbai in his forties, migrated to New Zealand in 1993; but unable to find a proper job in his field, he migrated to Australia in 1997. He was studying IT at Advance Technology in 2000, and this was how I met him. (62–63)

Jathavi, a Tamil IT graduate, went to Sydney in November 1999 through Piranavan. Dissatisfied with the situation in Australia, he went back to India in late 2001 to try to go to the United States: "At home you may not want to move. When abroad, you are forced to try this and try that. You become more ambitious." (90, 95)

Joseph, an IT professional in his early forties from Kerala, was among the earliest master's degree holders in computer science in India, and he also had a German MBA degree (1994). He was assigned to Sydney in 1998 by a Chennai-based software company but he resigned and set up E-Bet. (71, 75–76)

Kaaveri, a Tamil in her late thirties, migrated to Australia as a dependent. A graduate in mathematics, she provided tutorials to Indian school-children while working in small companies in Sydney. In 1999 she started an IT training course at home. I met her through her husband, Kalaimani. (161n4)

Kalaimani, a Tamil in his late forties, had studied engineering in Russia during 1981–83, sent by the Indian government. After working for two large national companies in India, Kalaimani migrated to Australia in 1993 and has worked as an IT contract consultant since. (89, 92)

Kana, a thirty-six-year-old Kamma businessman, told me immediately when we met for the first time in his Advance Knowledge that he had graduated from the Indian Institute of Management (IIM) in Bangalore, which was, according to him, the Indian equivalent to Harvard. Kana migrated to Australia as an independent skilled migrant in 1996, and with the capital brought from India, he started trading computer hardware from Taiwan to Australia and India. In 1998 Kana, Ravi, and Vijayarka set up Advance Technology, but the partnership soon broke up. Advance Technology was closed down in late 2003, and in 2004 Kana worked at an Australian food company as an IT specialist. Allegedly in huge debt, he was sharing a small flat with a few bachelor IT workers (he was divorced by that time). (60–61, 64, 71–72, 81–82, 94, 96–99)

Karthik, a Tamil IT professional in his early thirties, had dreamed of going abroad since childhood. His father, an engineer, applied to migrate to Australia in 1990 but failed. Karthik went to New Jersey through a body shop in 1994. After two months of benching, he changed his sponsorship to another Indian-run firm, despite the demand from the previous sponsor for a compensation of USD 20,000. Karthik explained: "I can't change to [be sponsored by a] big company. For small companies, you have to keep changing to know which one is good. Then you can stick to the [good] one and ask them to sponsor your green card." In 1995 the father's application for migration to Australia suddenly went through; Karthik visited Australia, arranged to marry a woman studying there, and became an Australian PR—all within one year. He went back to the United States in 1996 to apply for a green card, but by 1999 could not wait anymore and went to Australia. When I met him through Sivakrishnaviram, a remote relative of his, he planned to take a part-time MBA course and to gradually move to a managerial position. (53)

Kasi, an IT worker in his mid-twenties from Tamil Nadu, went to Sydney in September 2000 after being introduced to Piranavan by a body-shop operator in Chennai. After being benched for a few months, Piranavan persuaded him to return to India to "wait for opportunities." (96)

Kejal, chairman of G&J IT Solutions and originally from Mumbai, migrated to Perth in 1978 and then to Sydney in 1984. G&J's main business was Web site design and maintenance, particularly for "ethnic businesses" such as Indian and Chinese restaurants. (92)

Ken, a retired school teacher in his early sixties, was the first ever college graduate in his village in Tamil Nadu. He went to Ethiopia as a school manager in 1966 through application in response to a newspaper advertisement. Although no one in the village knew where "Ethiopia" was (Ken himself found out only after his arrival that the country was in fact poorer than India), leaving India per se was celebrated by the entire village. Ken got married as soon as the passport arrived. In the early 1970s, when the political situation in Ethiopia deteriorated, Ken applied for jobs in the United States, the United Kingdom, Canada, and Australia, and migrated to Australia in 1974. He then brought his wife in and had two children there. I met Ken in the Sri Venkateswara Temple where he was a volunteer overseeing the second phase of construction. (40)

Keshev, a Telugu chemical engineer in his mid-thirties, went to Sydney through the family-reunion program in 1999. He then joined Soul Networking, the largest body shop in Sydney and took an IT course at the same time. (78)

Kishore, Ashwin's brother and Uday's cousin. (39–40)

Kondepudi, a mechanical engineer from Karnataka, migrated to Australia through the family-reunion program in 1997 since his wife's brothers had migrated to Australia in the 1980s. Since 1995 Kondepudi has worked as a freelance consultant; at the same time he has registered his own service company and hired two workers. Despite the economic downturn, his business was doing well and by the end of 2001 he hired five workers, all paid according to the amount of work they were subcontracted to. (56)

Khrishna, a Kerala native in his late forties, went to Australia in 1991. He originally studied chemical engineering in a second-rank university because of his Brahmin background and the reservation system—"the worst thing in India" according to him. The oil crisis in the 1970s convinced the family that chemistry would not bring about any promising international job prospect, and Khrishna joined the Indian Institute of Sciences in Bangalore to become one of the earliest master's students in computer science. In Australia Khrishna was laid off after working at a local company for less than one year. Khrishna saw psychiatrists for one year, and his family avoided meeting any fellow Indian friends (it was particularly hurtful to him to have to withdraw his daughter from a private school and send her to a state one). After that Khrishna worked as a contract consultant, searching for jobs through agents. A subscriber of *Fortune* and *Harvard Business Review,* Khrishna was well versed in global business trends. When I met him in 2001, he was seeking senior positions in the Australian branches of large Indian IT companies—which was a fairly common choice for Indian IT professionals who had worked overseas for a long time. (53, 56, 104)

135

Lahudoss, a lecturer in an engineering college in Tamil Nadu, and a family friend of Piranavan, introduce a couple of his colleagues to Piranavan to be sent to Australia. (90)

Laxman, from Karnataka, was assigned to Australia by IBM India in 1998. He stayed on and became a PR. I met him at an IT workshop organized by Cisco Sydney in 2000. (159n2)

Leela, a Tamil Brahmin and IT entrepreneur, is Sivakrishnaviram's daughter-in-law. (57)

Madhu, a Kamma IT student from Guntur district of Andhra Pradesh, was studying in Melbourne, sponsored by his uncle. (37–38)

Mahathi, a undergraduate student in Bhimavaram town, West Godavari district, Andhra Pradesh, was working as a teach assistant at a home-based English-coaching class run by a high school English teacher. I found the class following its advertisements on electricity poles that claimed to provide training tailored for IT professionals and after-training placement advice, though the owner had yet to establish any contact with IT companies. Mahathi was taking an IT course at the same time. He planned to go abroad "at any cost." (127)

Mandu, from the Nair community in Kerala, is one of the most senior Indian IT professionals in Sydney. Graduated from Kerala University in 1969, Mandu joined an IBM training course and subsequently worked for IBM in Delhi. Shortly after he joined the company, the Indira Gandhi government required IBM India to sell part of its equities to Indian companies. IBM refused and closed down its Indian branch in May 1978. As a result, Mandu estimated, about five hundred former IBM employees emigrated after 1978, including thirty to Australia. He chose Australia instead of the United States and Canada because of its weather, and cricket. Mandu was now much closer to Keralite doctors than to IT workers partly because of golf, which he and the doctors played together, and young IT people had yet to learn. (not mentioned in text)

Mani Sandilya, a Tamil IT professional in his forties, migrated to Australia in May 1989, and has worked as a contract consultant since. He was keen on pulling strings between workers, body-shop operators and large Australian agents. In our first interview, he handed me a note: "Human relationship is the most important factor for every success." (76, 78)

Maruti, twenty-four years old, went to Sydney in 2000, sponsored by Advance Technology and introduced by Vand. Dissatisfied with the technology level in Australia and unable to find exciting opportunities in the United States, he returned to India in 2002 to run an IT consultancy with friends. (61)

Meena, a twenty-eight-year-old Telugu with a bachelor's degree in biology, went to Australia in 1996, one month after her marriage, following her husband, who was assigned to Sydney by American Express India. In Sydney Meena took a master's course in infomatics and then worked as a part-time contract worker after failing numerous applications for permanent positions. This, according to her, changed her approach to life completely, and made her more determined to go the United States: "Where is the real security from without money? If one year's change can makes five years' money, then it means five years' security." She further suggested that this is particularly true for women since it is difficult for them to have long IT careers anyway—she often studied and worked until two o'clock in the morning, after housework, to keep up with the demanding workload; such a schedule is not sustainable. (104)

Nadarajah, a Tamil in his fifties from the Chettiar caste, worked beginning in 1985, at the Center for Development of Telematics (C-DOT) in New Delhi; he was one of its earliest employees. C-DOT was set up at the suggestion of a prominent U.S.-based NRI, Mr. Sam Pitroda (who was called Uncle Sam in India), with support from Rajiv Gandhi; he hoped to build a telecommunication exchange system covering most parts of India. After 1988, however, alleged conflicts between Sam Pitroda and some federal ministers escalated, and employees gradually left. Of the 1,200 employees in 1990, as many as 1,000 were estimated to have emigrated over the years; and about 30 moved to Australia during the six months from late 1998 to early 1990 alone. Nadarajah migrated to Australia as a skilled immigrant in 1989 because his brother, Chaya's school friend, was doing a Ph.D. there (but later returned to India). Nadarajah was laid off in 1991 during the recession and became a contract consultant. In 1999 he started a home-based training course to help benched Y2K programmers learn the then much demanded Java skills; this became a formal business, TrueInfo, in 2000, as a joint venture with Chaya. Nadarajah's two children were taught in Hindi and English in Delhi, and he was so excited to learn about the Tamil school in Sydney that he went there the first weekend after his arrival. Since then he has been one of the most active figures in the community. (66–67, 88)

Nambi, a young Tamil from Colombo, Sri Lanka, studied in Chennai in 1994, returned to Colombo in 1996, and moved to Australia in 1998 on a 457 visa. (51)

Narendra, a Telugu IT professional, went to Sydney through Piranavan in early 2001. He was benched for five months, and in the end found a job in a construction company in Sydney. (38, 84–85 [box 1])

Navin, an IT professional in his early forties, was assigned to Australia in 1996 by Tata Consultancy Services. He became an Australian PR in 1997, traveled to the United States, and then returned to Australia to settle down in 1998. (101)

Panika, twenty-seven years old, previously a journalist in Kolkota (Calcutta), went to Australia as an IT student in 1996. While still a student, he set up the Web site Fiji Times and the portals Indian Times and Indian Links as vehicles for business and marital advertisements in the community in Sydney. He hired six freelance reporters to contribute stories as well as bring in advertisements. When I got in touch with him through his Web sites, he was planning to go to Vancouver through a body shop there. (18)

Piranavan, forty-seven years old, is of Tamil origin but grew up in Hyderabad, as his father had been working for the Andhra Pradesh state electricity board. After graduating from Hyderabad University, he worked in a large national company in Chennai and was then recruited by an automobile company in Dubai in 1981. Piranavan set up the entire computer section for the company and recruited forty IT workers, mostly from India, but also from Pakistan and the Philippines. In 1985 he applied to migrate to Australia through a job advertisement posted by the Australian embassy in Dubai. He set up the consultancy Osin System in 1991 and started body shopping in 1999. Piranavan's brother migrated to Australia in 1990 from Saudi Arabia, and also helped Piranavan with software development in his spare time. The two families had always lived in the same suburb and often organized large *puja* together. Although not committee members of any association, the brothers had been active in helping organize community activities. (56–58, 62, 68, 71, 75–76, 79, 88–90, 95–96, 98)

Prakash, a Telugu IT worker, went to Sydney in 2000 sponsored by Advance Technology. He left for New Jersey on an H-1B visa in March 2001, but returned to Australia in June of the same year, still with Kana's sponsorship. (99)

Puli Reddy is the owner and operator of the body shop Puli Reddy Consultancy, part of the Puli Reddy Group registered in Hyderabad. Puli Reddy had the reputation of "spending money like an Aussie," and he was constantly on the move between Sydney, Hyderabad and various cities in the United States: to seek new businesses, to enjoy life, and, I was told, to avoid his wife. By 2005 he had given up his body-shopping business and gone back to Andhra Pradesh to join his father's construction business. (46, 48, 77, 79, 93)

Raj, a Telugu electronic engineer in his forties, went to Australia in 1991 through the family-reunion scheme (his wife had two brothers in Australia). He set up his Raj Electronic Consultancy in 1997. I met him at the Sydney Telugu school. (57)

Rajalaxmi, an electronic engineer by qualification, and in her early forties, went to Australia with her husband, an IT professional, in 1992, and worked in the Tax Reform Office of the federal government. She was involved in a body-shopping business operated by her mother and brother. In the early 1990s, her mother attended a seminar held by NRIs in New Delhi whence she learned about the Y2K problems. In 1992 she visited Rajalaxmi's brother, an IIT graduate, in the United States. To make their mother feel less lonely, her brother organized parties inviting over his colleagues and friends; at these parties the mother picked up leads on possible labor demands at different companies, and sent workers to five companies within three months of returning to India. From then on, the mother and brother worked together to send over ten workers to the United States every year and more than one hundred workers a month at peak times after 1995. In 1998 Rajalaxmi helped her mother bring workers to Australia, and she took a leave to study IT in India for four months so as to have a better understanding of the market and technologies. But Rajalaxmi never developed the business aggressively due to her husband's objection. (103)

Rajan, previously a lecturer in an engineering college in Tamil Nadu, went to Sydney with Piranavan's sponsorship in November 1999. After being benched for about two months, he was placed through E-Bet, a body shop run by Joseph, Piranavan's friend; he subsequently worked on Piranavan's software-development projects. He became a PR in 2002. Rajan was introduced to me by Joseph, and he was instrumental in bringing me in contact with other workers sponsored by Piranavan. (xvi, 90, 93, 95–96)

Rajashekhar Reddy was the founder and principal of I-Logic, an IT training institute in Hyderabad. He had previously run a software consultancy, but a series of failures in finalizing deals with government departments and companies convinced him that as a Reddy he could not compete with Kammas in IT services (on more than one occasion the prospective client changed his mind upon learning his surname), and thus set up the training institute. (29)

Rajiv was manager general of G&J IT Solutions, which was founded by his father, Kejal, and migrated to Perth in 1978 at the age of seven. He returned to India for schooling for four years, and thus had more friends in India than in Australia. His wife lives in Sydney and Mumbai alternately. (92)

Rama, an Indian-Fijian in his late thirties, migrated to Australia with his wife, a teacher, in 1992 through the independent migration scheme. Previously a customs officer in Fiji, he ran four businesses in Sydney: a spice center, a travel agency, a car rental service, and a computer exporting business (to Fiji). In the wake of the coup in Fiji in May 2000 (which overthrew Mahendra Chaudry, then prime minister, and of Indian origin), Rama called all India-related organizations, including Hindu temples and mosques, and Australian politicians (from both the ruling and opposition parties) to hold a gathering to condemn the coup. Most Indian associations, according to Rama, were lukewarm in response. Rama explained that the Indians from India were "sympathetic [with Indian-Fijians because] we are all Indians . . ."; but they were not enthusiastic because "they are professionals, they don't care about politics." (57)

Rama Chandria, a Kamma from the Guntur district of Andhra Pradesh, unemployed in Hyderabad, was exploring the body-shopping business with Ravinder in 2001. (42, 46)

Rambabu, from a backward caste in Tanuku town, West Godavari district, went to London to study theology in 1995 and returned to his hometown in 1999 and became a migration agent. (52)

Rangarajan, a Tamil IT professional, worked in Singapore for more than ten years. (65–67)

Rani, twenty-three years old, a master's student in Bhimavaram, West Godavari, wanted to become a college teacher in her hometown. (128)

Rao Prabhala, a contract consultant in his forties, went to Wellington in 1992 along with forty other Indians, when recruited by Computer Science New Zealand, an IT firm. He migrated to Australia in 1996 from New Zealand. (78)

Ravi, a Telugu in his late thirties, went to Australia in 1994 as a master's student. He graduated and became a PR in 1996. He returned to India immediately, worked for a few private companies, and went to Australia again in late 1997 as a representative of Neko, a large Hyderabad-based IT company. But Ravi soon broke away from Neko and joined Kana to set up Advance Technology. The two separated less than one year later, and Ravi began running his own body-shopping business. By 2004 Ravi had shifted his business focus to the trading of Telugu and Tamil movies and TV dramas in Australia and Southeast Asia. (60–61, 64, 68, 72, 78–79, 82)

Ravi Kolluru, an IT professional in his early thirties, graduated from the MCA program at Andhra University in 1997. He first worked in a medium-sized IT company in Hyderabad, but he thought the company was "too slow and had no offer to go overseas," so he joined another company in Bangalore, and subsequently went to Australia through a body shop in July 2000. He returned to Hyderabad in 2001 partly because of the depreciating Australian dollar. I met him in Hyderabad through his uncle, an academic at the University of Hyderabad. (40–41)

Ravichandary, thirty-eight years old, was one of the first workers at Advance Technology whom I got to know, as he was working as a receptionist. Ravichandary got to know Uday when they were both "free men" (Uday's term for the unemployed) roaming in Hyderabad, and Uday introduced him to Advance Technology in late 2000. He returned to Hyderabad in late 2001 and then went to Dubai in 2002. (96–97, 113)

Ravinder, born in 1971 to a business family that owned a small pharmacy in Vishakhapatnam, was a typical IT transnational lumpen bourgeoisie. Having finished his master's program in 1997, Ravinder went to Hyderabad looking for jobs, where he fell in love with a doctor. The doctor had married an NRI in the United States and had a daughter, but had found out that the man had been married in the United States. She took the NRI to court and was said to be compensated INR 10 lakhs, but this also caused her to be cast out by her family. With her support, Ravinder went to Singapore twice within three months in 1998 and 1999. Although failing to find a job, Ravinder met a few Indians whom he called "waiters"—waiting for jobs just like him—in Singapore. Back to India, using the connection with his school friends in New York and Chicago, he sent two "waiters" from Singapore to the United States, and earned INR 8,000 from the deal. At the same time, Ravindar got in touch with Puli Reddy and went to Australia in 2000, while his girlfriend went to Libya through a government program. Ravinder returned to India in early 2001, and went to Australia again in December 2001 but was deported from the Sydney airport. In 2003 he went the United States with friends' support. (41–42, 46, 48, 92–93)

Rejeshekhar, a twenty-five-year-old Telugu IT professional, went to Kuala Lumpur in 1998 with his brother-in-law, sponsored by his friend. I got in touch with him through the Internet, and he and his brother-in-law were my key informants in Kuala Lumpur in June 2001. (46, 106, 120)

Remash, a Telugu food specialist, went to New Zealand as an independent skilled migrant in 1995. He soon discovered that it was almost a norm for Indian professionals there to study IT—partly in order to move on to Australia. In Auckland Remash worked every day from 6 P.M. to 6 A.M. in a deli company, slept from 6 A.M. to the noon, then studied IT in a New Zealand institute from 1:30 P.M. to 4:30 P.M., after which he resumed his day. He was laid off in 1998. He took more IT courses and did four projects for small software consultancies. Remash migrated to Australia in 2000, but still could not find a job; he set up WinWin Recruiter in late 2000. (58, 78)

Sai, a Kamma Telugu in his early forties, from a rich business family, and a mechanical engineer by education, had done business in Chennai, but was forced to return to Hyderabad by his father and sister, who were trying to get him married before he got too old. While little progress was made with marriage, Sai quickly realized the great commercial potential in "helping IT guys go abroad" from Hyderabad, and thus set up M-Station, though he had little interest in IT. (45, 48–49)

Sam, forty-six years old, went to Australia sponsored by Chaya in February 2000. After graduating in 1977 and working for government departments in various places in India as an electronic engineer, Sam studied computer science for two years in IIT (1982–84) under government sponsorship. In 1998 he quit the government job, because of the low pay, and joined a U.S. company in India. Two years later he decided to go abroad because "if you don't understand how the international market operates, you can't be promoted." At this point Sam met his old friend Chitra, who used to be a neighbor of Chaya's parents in Chennai. Another neighbor of Chitra's, working in NIIT, could not wait to be assigned overseas (NIIT normally requires a new employee to work in India for three years first) and had himself sent to Australia through Chaya. Chitra thus learned of Chaya's business and introduced Sam to Chaya. Disappointed with the situation in Australia, Sam was planning to return to India when I met him in an Indian grocery store. (91)

Samy, in his midthirties, from New Delhi, went to Australia as an MBA student in 1992. On graduation in 1994 he returned to Delhi to join his father's trading business, and later set up his IT training institute. In 2000 he went to Australia as a business migrant with an investment of AUD 250,000, and the owner of Soul Networking in Sydney, a close friend of his, asked him to join his business. He set up Glogo Consultancy as a subsidiary of Soul Networking nominally but he operated it independently (by doing so both Soul Networking and Glogo benefited in regard to taxation and marketing). (51, 74, 91, 103)

Senthil, twenty-five years old, a Telugu, went to Australia in 1998 as a student, converted his student visa to 457 in 1999, and became a PR in 2000. In late 2000 he joined a small IT agent run by a Frenchman but left soon afterward. (60, 103)

Sharma, a Telugu in his early forties, started his career in the early 1980s with Patni Computer Systems (PCS), a well-known IT company set up in Mumbai by a returned NRI from the United States. In 1983 he joined Minicomp, another firm set up by two NRIs. In 1987 he went to Saudi Arabia, and in 1995 migrated to Australia. He did not go to the United States because he disliked "lies": during the first Gulf War, he saw CNN broadcast the images of piles of dead seagulls as evidence for the environmental catastrophes resulting from Iraq's war tactic of burning oil wells; he believed these images had been manipulated based on his firsthand observation. In Sydney Sharma was an active member of the Telugu community, and we met at a function. (91, 103)

Singh, a Sikh in his early thirties, went to Australia in 1994 as a skilled immigrant. He had been working as a civil engineer, and was taking an IT course at Advance Technology in 2001. (63)

Sitaram Reddy, a Telugu IT professional in his early forties, migrated to Australia in the late 1980s as an independent migrant, and started Softworld Consultancy in 1993 with his wife when both were laid off. (55–56, 58)

Siva, twenty-four years old, from Vishakhapatnam, went to Australia in 1997 as a student of Queensland University. (101)

Siva Prasad Rao, a Telugu journalist in his early forties, migrated to Australia in the 1980s as a refugee. Unable to find a job in journalism, he started a home-based IT training course in 1999, and subsequently ventured into body shopping. I met him through an advertisement for this training course in an Indian business directory in Sydney. (64)

Sivakrishnaviram, an Iyer Brahmin from Tamil Nadu in his late sixties, studied and subsequently worked in the Indian Statistic Institute in Mumbai throughout the 1970s. In 1982 he quit the job and set up a computer consultancy in Mumbai. The business did not work out well; in 1985 he and his family migrated to Australia through the family-reunion scheme, as his elder brother, a doctor, had settled in Australia in 1969 after studying in the United Kingdom. Sivakrishnaviram has worked as a freelance IT consultant since. His two sons, one daughter, and two daughters-in-law are all IT professionals; even their everyday life has been colonized by IT language—I was once asked to help his daughter "download" groceries from the car. In Sivakrishnaviram's

143

extended family, about ten families were in the United States and five in the Middle East; they met up once every two or three years in India, mainly at relatives' weddings. (58)

Sree Kumar, an IT worker from Kerala in his late twenties, went to Australia in 1999 sponsored by Advance Technology and became a PR in 2001. (54, 94, 98)

Srina and **Shyla,** a Kannada couple, were in their early forties when I met them. Srina migrated to Australia in 1998, which was in part a result of his studying at the National Technical Training Foundation (NTTF) in Bangalore in the 1980s. NTTF had been set up by a Swiss missionary to provide free vocational training to poor youth in India. The institute manager was said to have embezzled donations from Switzerland into horse racing, and the school, lacking funding, had to recruit self-financing students from middle-class families. Fifteen of the twenty-five students in Srina's class had gone overseas by 2001, including eight who migrated to Australia. Although a mechanical engineer who was working in a factory in Sydney, Srina was actively exploring the body-shopping business in 2001. Shyla worked part-time in a beauty salon. They had two young children. (90, 101)

Subar went to Sydney in 1999 and registered two companies in 2001. He later shifted his business focus from normal body shopping to helping Indian students in Australia convert to 457 visas, which according to him was more profitable. (8)

Sudheer, an IT professional in his late forties working at a state government department in Orisa, went to Singapore in 1998, then to Australia through Advance Technology in 1999. He went back to India three months later. (97–98)

Suman, a twenty-six-year-old Telugu IT worker, went to Australia sponsored by Piranavan in 1999. He was completely shaved when I met him in a temple in Sydney: he had just offered all his hair to Sai Baba in India. He showed me his two finger rings, which were given to him by his parents upon his birth, one for academic success and the other for wealth. Suman chose IT as his career because, as he said, it combined the two. He introduced me to his flatmates who were also sponsored by Piranavan. (88)

Sumanth, a Kamma born in 1981, went to Sydney in 2000 as a business-college student immediately after completing high school. Sumanth's father had died in a car accident in 1999. Although the family survived well economically with cotton mills, a paddy field in the countryside, and shares in a transportation business in Hyderabad (a business strong-hold of Kamma's in Andhra Pradesh), the accident prompted the brothers to go overseas quickly to seek new income sources for the family. Sumanth told me that he "must go to London" at our first meeting, and he successfully did so in 2002. His older brother, an IT professional, left India in 2000 for Singapore, and subsequently went to Malaysia and Bangkok looking for jobs. (164n8)

Susai, a twenty-six-year-old Telugu IT worker, went to Australia sponsored by Piranavan in 1999, and returned to India in mid-2001. (88, 90, 95)

Syed, a Muslim IT professional, had worked in Dubai, returned to Hyderabad for family reasons in 1997, and set up the body shop Zentech. (49–50, 52)

Tambi, a Tamil in his forties from Sri Lanka, ran a grocery in Ashfield, western Sydney. (117–19)

Uday was born in 1971 in Vishakhapatnam; his father was a superintendent engineer working for the district Public Works Department, and his mother was a housewife. While Uday was searching for jobs after completing his master's program at Andhra University in 1997, one of his cousins went to the United Kingdom, one to the United States (Ashwin), and another (Kishore) to Singapore. Making things worse, Uday's parents' house was close to the Vishakhapatnam airport, and the cousins and even their friends often stopped over on the way in or out, talking loudly about their overseas experiences deep into the night. In 1998 Uday volunteered at a local IT company in Hyderabad, where he learned about Singapore from a fellow employee. He went Singapore five months later on a tourist visa. Determined to not contact his cousin (Kishore) there, he put up in a hostel in "little India" but could not find an IT company. Luck came his way the day before his visa expired. He ran into an Indian at a bus stop who gave him the address of a small company run by an Indian and a Chinese that claimed to be a branch of a Beijing-based corporation. The company extended Uday's visa instantly. This taught him the golden law in judging a body shop: "Just look at how fast they can deal with the immigration department." The company placed Uday in three jobs until 1999 when Kana and Ravi of Advance Technology went to Singapore to recruit. Since most of his friends in Singapore—about twenty Indian IT professionals—chose to go to North America, Uday thought Australia would be less competitive and that he would be treated better there—a decision that paid off: he became an Australian PR in June 2000, well before any friend in the United States had gotten a green card.

With permanent residency in hand, Uday left Advance Technology and found a long-term contract job at an Internet game company through an Australian agent. In 2001 Uday married a girl from his hometown. He and Ashok bought two flats jointly in Sydney; they lived in one of them and rented out the other until Ashok got married and moved there in 2003. Uday's older sister was married to an engineer and lives in Vishakhapatnam, where their parents still reside. (xvi, 23, 25, 28, 38–40, 46, 54, 61, 89–90, 92–95, 97–99, 101–3, 105–6, 119, 163n4, 164nn4 and 8)

Umesh, a twenty-seven-year-old Hyderabadi, went to Sydney in November 1999, sponsored by Piranavan. From the age of fourteen, he had written letters to the BBC and music clubs in the United Kingdom and had always hung out with "open, ambitious, and cosmopolitan" friends. He did not go overseas earlier because, according to him, he was in the wrong occupation, chemical engineering; and he started studying IT only in 1998. After arriving in Sydney he was sent to another body shop to help out with software development without pay, and was then placed in May 2000. Umesh became a PR in mid-2001 and left Piranavan soon afterward. (89–90, 95)

Vand, a twenty-three-year-old Brahmin originally from Bihar, went to Australia in 1999 sponsored by Advance Technology. Vand had wanted to become an Indian administrative servant who would "sit next to the chief minister," and for this dream he had practiced scouts and Chinese martial arts since high school. But his father did not allow him to study arts, and instead spent INR 150,000 to send him to Vishakhapatnam to study IT—Bihar did not have good technology colleges at that time. Vand quit his job at Advance Technology shortly before the body shop was closed down in 2003. In 2005 he was working as a senior program analyst in an Australian IT company. (61, 96–97, 99)

Vasan, twenty-six years old, majored in English literature at university, studied IT in training institutes after graduation, and was running a franchise of Visu Consultancy in Bhimavaram, West Godavari. Vasan said he would "definitely" stay at home although he was keen to send others abroad. (126–27)

Venkate, a Hyderabadi in his early thirties, worked as a salesman in a life insurance company in Chennai and Hyderabad for three years until 1998, when he took a six-month leave to study SAP at the training institute cum body shop Global Intelligence in Hyderabad. He explained why he had decided to shift to IT: "Marketing is like cricket. You have to play together with your team mate. You spent a lot of energy but no much recognition. IT is like tennis, it's *your* show." After failing his visa application to the United States, he went to Australia in 1999 through Advance Technology. In November 2000 he left Advance Technology upon becoming a PR. Although not a close friend to me, Venkate was very helpful in clarifying how body shops handled policy issues based on his extensive consultations with Australian immigration solicitors, taxation consultants, and IT placement agents while preparing his application for permanent residency. Venkate once joined an Indian association, but found it unhelpful: "Young single fellows don't have the time for this [organizing associations], but when you are established and become a leader, you forget what you had gone through. Every person has his private agenda." (18, 81, 98, 101)

Venu, a Telugu 457 visa holder, went to Sydney in 1998 through a body shop. Outgoing and a beer lover, he soon became a friend to Puli Reddy from whom he learned much about the business of body shopping, and he was exploring to set up his own business in 2000. (73–74)

Venugopal, twenty-four years old, a Kamma Telugu, went to Australia in 1999 on a 457 visa. (100–101)

Venush, a twenty-seven-year-old Telugu IT worker, went to Australia sponsored by Piranavan in 1999. Venush always had good news for workers on the bench—citing his brother who was working in the Chennai Stock Exchange, he projected that the IT market would recover soon. (xvi, 89–90, 95)

Vijayarka, a Telugu in his thirties, a friend of Kana's, went from Vishakhapatnam to Sydney in 1998 to join Kana's business. But they soon broke up and he had to ask a Telugu who was running a furniture factory to be his visa sponsor. Vijayarka lived in the same apartment building as Uday and I, but he avoided us—he was visibly depressed due to being jobless for half a year. In 2004 he was hired by a large bank in Sydney. (61)

Vikram, a twenty-eight-year-old Telugu IT professional, went to Australia in 1999 on a 457 visa. Always ambitious, he applied to go to the United States for his "eighteen years" when he finished "sixteen years" in Hyderabad in 1995 (according to Vikram, sixteen years and eighteen years are American phrases for undergraduate and master's courses). Failing, he worked in various IT multinationals in India. In 1998 he went to Singapore through a body shop and worked at Exxon Mobil. His wife joined him and did a correspondence master's course in commerce with a college in India. Although he thought Singapore a good place to start a career, he did not plan to stay there for long: cars and private houses were too expensive. He has since gone to Australia through a body shop in Singapore. (59)

Vinnie, a Hyderabadi IT professional in his midthirties, went to Australia in the early 1990s as one of the earliest students recruited through private agents. Vinnie started working part-time a year later and became a PR in 1994. After working at five jobs over four years, Vinnie registered a series of companies: Achieve, Vision, and ComLink. The receptionist at ComLink was an IIT graduate in his early forties who went to Australia through another body shop and had been benched for a few months. Vinnie offered him this job simply because he was an IIT graduate. Vinnie was active in the Telugu community, and a regular organizer of cricket matches among Indian IT people. (24, 64, 67)

Vishnu, twenty-five years old, Gopal's flatmate in Hyderabad, had just quit his job as a medical transcriptionist when I met him. (161n5)

Introduction

1. A projection by the Information Technology Association of America (ITAA) in 2001, cited in *San Jose Mercury News,* "Demand for IT Workers Is Down 44 Per Cent," April 2, 2001.

2. *CBS Marketwatch,* "Landscape Shifts for Laid-Off Foreigners," August 31, 2001.

3. *Los Angeles Times,* "U.S. Tech Firms Abusing Visa Program, Critics Say," November 21, 2001.

4. This estimate is based on the main media sources that I monitored throughout my research during 2000–2001, which included the *Times of India,* particularly its Hyderabad edition; *Computer Today* (India); and *Sydney Morning Herald.* The mailing lists *CISNEWS* of the Center for Immigration Studies based in Washington and the *Age Discrimination/H-1B E-Newsletter* compiled by Mr. Norm Matloff, professor of computer science at University of California, Davis, provided comprehensive leads to other media reports.

5. *Times,* "Americans Pull the Plug on Indian Computer Whiz-Kids," May 15, 2001.

6. I use the term New Economy in the conventional sense, referring to a knowledge-based economy organized around the flexible production of goods and services, and said to require new (deregulated) trade, tax, and employment policies.

7. Polanyi (1944; 1957a; see also Carrier 1998a; 1998b) was perhaps the first to explicate the trend of abstraction of capitalism, though earlier observations had pointed to the same, for example Marx's "commodification" and "alienation," and Weber's "rationalization."

8. I use the term "neoclassical" to refer to the mainstream economics theories

since the 1970s, and "neoliberal" for the political ideology that believes in a free market and minimum government intervention. Obviously the two are interrelated.

9. All too often, the current usage of "embeddedness" refers to "disembeddedness" following Polanyi, a slippage associated with Granovetter's (1985) now near classic article. According to Polanyi, "embeddedness" is the condition wherein material production and exchange are enmeshed in a priori social relations that are independent from these economic activities and therefore remain intact when these activities cease. For instance, "embeddedness" could hardly be used to describe two Indian immigrants in Sydney coming together to do business when it is clear that their "co-ethnic" relationship may well terminate once their business together ends. Sometimes, as seen in the more recent anthropological literature on migration, embeddedness is so vaguely defined as to say little more than "context" and/or "networks" (e.g., Kloosterman and Rath 2001).

10. *Mercury News,* "Labor Contractors of All Stripes Abound in Bay Area and the Nation," November 19, 2000.

11. My estimate is based on the following data released by the Immigration and Naturalization Services (INS), United States (2000), for the period of May 1998–July 1999: (1) 63,900 Indians were issued H-1B visas (47.5 percent of the total issued), the long-term temporary work visa that most body shops used to sponsor IT workers to the United States; (2) Indians made up 74 percent of all the IT professionals on H-1B visas; and (3) more than 60 percent of all H-1B visa holders were IT professionals. At the same time, according to a media report, there were an estimated 250,000 Indians on H-1B visas by 2001 (*San Jose Mercury News,* "Back to India for Tech Worker," February 3, 2002).

12. In the three consecutive years from 1980, Tata Infotech, another IT firm in the Tata Group, India's largest industrial conglomerate, recruited about 100, 300, and 500 graduates a year to be sent out on its growing on-site-service business overseas. Quite a few of these graduates, usually of the elite state institutes, founded the earliest body shops in the United States capitalizing on their client connections (interview with a former employee of Tata Infotech, August 4, 2000, Sydney).

13. Partly because of this, body-shop operators in both India and Australia normally referred to themselves as "IT consultancies" while IT workers and other companies (particularly large Australian IT placement agents) would often call them "body shops."

14. There is a large amount of literature along this line. For instance, it has been repeatedly stressed that one would be excommunicated by the community in case of failing to honor a debt, and therefore everyone is pressured to comply with the rules, and thus, economic and social order is sustained (e.g., Letyon 1970; Wallman 1975; Herman 1979; Kosmin 1979). A methodological problem with this approach is that it attempts to explain individual behaviors by what are in fact emergent consequences of these behaviors. A more important question may concern how these collective social forces, such as enforceable trust, are achieved through daily interaction based on individual behavior, rather than how people behave with norms and structure that have already been well established.

15. Classical social scientists such as Tocqueville, Durkheim, and Mauss suggested that the individual in the modern times is disembedded from ascribed in-

stitutions such as lineages, and thus becomes a self-determining entity. Postmodernist individualism, according to theorists such as Giddens and Beck, is linked to consumerism, the loss of class commitment, reflexive agency, etc.

16. Mastech was set up in 1988 by two Indian graduates in Pittsburgh and initially specialized in recruiting Indian IT workers for "off-Broadway" cities in central or southern parts of the country where few IT professionals seemed willing to go (most were concentrated in cities like Boston, Los Angeles, New York, and Houston). (Interviews with Nandan Desai, manager, Mastech Australia, August 1, 2000 and February 4, 2001, Sydney.) Unsurprisingly Mastech was a top employer of H-1B visa holders in the United States—the fourthmost in 2000 (*eBusiness*, "More Bangalore for Their Bucks," April 9, 2001).

17. By replacing the term "*inter*-national" with "*trans*-national," the notion of transnationalism argues that examining the relations between nations with clear-cut boundaries is not sufficient to understand the current world. Instead, transnational activities have obtained their own momentum and autonomy and thus constitute a new research field in themselves (Sklair 1995; Portes et al., 1999; Castles 2001). It is important, however, to bear in mind that, according to the United Nations, only 2.9 percent of the world's population live outside of their nations of birth in 2000 (Population Division 2003). Indeed, as Hirst and Thompson (1996, 161) remind us, immigration today is more difficult than it was one hundred years ago. The overwhelming majority of the world's population has no regular access to telephones, let alone Internet or fax. In terms of the flow of goods and capital, a substantial part of world trade is conducted as intrafirm transactions, and the major part of direct foreign investment goes to corporate mergers and acquisitions (Robinson and Harris 2000); as much as two thirds of direct foreign investment from U.S.-domiciled corporations in the 1990s was actually in their foreign earnings and unrepatriated profit (Sklair 2001, 84).

Chapter 1 The Global Niche for Body Shopping

1. Interview with Terry Porter, senior manager, ICON Recruitment Pty. Ltd., May 11, 2000, Sydney.

2. Internet technologies represent a new epoch in the evolution of computer science. Between the 1940s (when the first modern computer was invented) and the 1970s, the computer was developed and sold mainly as a machine (hardware), and software was treated as supplementary. In 1969 IBM launched a software division and for the first time customers were charged separately for hardware and software (Heeks 1996, 108); and with the advent of personal computers, software technologies became the focus of the industry. By the mid-1980s, software accounted for about 80 percent of all investment in computer systems (Eischen 2000, 3). The software business includes software package development (e.g., developing Windows XP) and software services (customizing, implementing, and maintaining software programs). Software package development is a highly profitable but capital-intensive and high-risk business; by comparison, software service is labor intensive, less profitable but stable. Software services comprised over

60 percent of the world software market by the end of the 1990s (Eischen 2000, 34), valued at USD 300 billion a year (*Far East Economic Review,* "Second Take-Off," July 20, 2000: 53–54). In India, the revenue generated from body shopping is counted as part of software services.

3. E-commerce in the broad sense includes all transactions conducted through electronic means, including the Internet, telephone, fax, etc. But Internet commerce is the main part of e-commerce.

4. A White House document (1997) stated that "[f]or over 50 years, nations have negotiated tariff reductions. . . . [T]he internet is truly a global medium, it makes little sense to introduce tariffs on goods and services delivered over the internet. . . . [The United States] will advocate in the World Trade Organization (WTO) and other appropriate international fora that the internet be declared a tariff-free environment. This principle should be established quickly before nations impose tariffs and before vested interests form to protect those tariffs."

5. *Deccan Chronicle,* "1999: Year of Business on Net," December 26, 1999. The term "Net year" was also used as a time scale—1 Net year equals 1.7 calendar months, a reflection of the short business cycle and high speed of change in the IT industry.

6. Venture capital is also called "risk capital" because in general only less than 10 percent of the investments in the IT sector turned out to be profitable in the 1990s, though the returns from successful deals were so high that losses could be written off. For example, Benchmark Capital, a prominent venture-capital fund in the United States, turned USD 6.5 million into 4 billion in 1999 by investing in eBay, an e-commerce firm (*Far East Economic Review,* "A New Benchmark," July 13, 2000). During the slowdown, some venture capitalists purchased failing dot-coms at rock-bottom prices hoping for a turnaround in the future, and were thus called "vulture capitalists."

7. The term "soft money," at one time referring to unlawful election campaign donations or one-off grants as opposed to recurring budgets in the United States, became associated with party contributions from software corporations. For example, Michigan senator Spencer Abraham, a leading sponsor of a bill to increase the number of H-1B visas, reportedly received USD 270,000 "soft money" between October 1, 1999, and February 29, 2000 (*Detroit News,* "High-Tech Firms Help Abraham: Companies That Would Benefit from His Bill Give to His Campaign," September 5, 2000).

8. In the United States, H-1B workers were widely reported to receive about 25 percent less than the market average and had no compensation for overtime (e.g., see *Asia Week,* "Contract Labor Program by Any Name Hurts All Workers," August 4–10, 2000). Furthermore, as wages for IT workers rose only a little more rapidly than for other highly educated workers in the United States in the 1990s (Appelbaum and Roe 2001), it was suggested that temporary workers were simply replacing permanent employees. In 1995 a large New York insurance company laid off 250 computer programmers and filled their jobs with lower-waged H-1B workers from India (*Washington Post,* "White-Collar Visas: Back Door for Cheap Labor?" October 21, 1995); and Dun and Bradstreet, an IT firm, was said to assign local programmers to train their H-1B replacements (Matloff, *Age Discrimination/H-1B/L-1 E-newsletter* circulated in 2001).

9. For how ITTA projected skills shortage, see *Times of India, Hyderabad,* "IT Jobs Are Here Again!" July 9, 2001.

10. *Information Age,* "Addressing the Skills Gap," February/March 2001: 24–28.

11. For more on official rhetoric, see for example Alston et al., 2001. Among the successes claimed were that between 1993 and 1998 about 150 IT or IT-related international companies established their Asia Pacific regional headquarters in Australia (Department of Industry, Science and Tourism 1998).

12. *Washington Post,* "Gates Assails Bill to Curb Immigration," November 29, 1995.

13. In the United States, for example, Intel continued headhunting workers in 2001 while paring 5,000 jobs; at the same time Cisco Systems reiterated that a high demand for skilled immigrants persisted while trimming up to 5,000 jobs (*USA Today,* "Once-High Demand for Visas for Tech Workers Slides," March 21, 2001).

14. The total amount of venture-capital deals in IT start-ups in the United States in the second quarter of 2001 dropped to USD 3.14 billion from USD 9.28 billion the year before (*Times of India, Hyderabad,* "VCs Backing Off after Tech Burst," August 16, 2001).

15. *Times of India, Hyderabad,* "IT Jobs Are Here Again!" July 9, 2001.

16. *eWEEK,* "Testing the New IT Job Pool," May 21, 2001.

17. *Times of India,* "Dotcoms See Hope Despite Gloom," December 30, 2001.

18. The constraints of the national accreditation procedure are sometimes reinforced by national professional bodies to curb immigration and thus preserve high wage levels and other career privileges (see Iredale 1998). By comparison, membership in the Association of Professional Engineers, Scientists and Managers (APESM) and the Australian Computer Society, the two major IT professionals' organizations in Australia, was open to anyone working in or even learning IT—in fact, many of those in decision-making positions in the associations had no IT qualifications themselves. According to the 1996 census of Australia, only 52 percent of employed computing professionals held degree-level qualifications, among whom less than half had computing credentials as their highest qualifications (cited in Birrell et al., 1999, 67–68). A survey in 2000 conducted by the IT&T Skills Taskforce, an industry-led team working closely with the federal government in Australia, reported that only 45 percent of IT employees had the equivalent of university degrees (*Information Age,* "Addressing the Skills Gap," February/March 2001:25).

19. Information technology corporations also extended the function of the certifications to serve their global business strategies. For example, Cisco encouraged Cisco-certified engineers to become its "premier partners" by generating a minimum amount of revenue selling Cisco products or providing Cisco services. In return Cisco alerted the premier partners to Cisco-related business opportunities. In effect, the scheme provided incentives for certified engineers to freelance as Cisco salespersons or service deliverers. When I was in Sydney in 2000 and 2001, Cisco organized monthly meetings that brought major Cisco users and potential clients together with premier partners.

20. As a mysterious figure who lived at the beginning of the eighteenth century in Shirdi village of Maharashtra, India, Shri Sai Baba commands much popularity

among Hindus, at home and abroad. Shri Sai Baba should not be confused with Sathya Sai Baba, the big-haired guru residing in Andhra Pradesh who claims to possess godly power and enjoys a global following, which includes countless Indian IT professionals overseas.

21. Interview with a former Indian employee of St. George's Bank, May 6, 2000, Sydney.

22. DIMA News Room, "Skilled Migration Changes to Boost Economy," August 27, 1998.

23. *Migration News*, "Australia: Asylum Seekers," 8, no. 11 (November 2001). As hinted in the minister's calculation, keeping asylum seekers and other undesirable migrants away, concurrent with attracting skilled migrants, is just another manifestation of the rationalization of immigration policy. This was dramatically encapsulated in the *MV Tampa* event of August 2001. *MV Tampa*, the Norwegian vessel that rescued 433 asylum seekers from their sinking boat in international waters between Australia and Indonesia, was ordered by the Australian authorities not to enter its waters. When the captain refused to alter his course on humanitarian grounds, the Australian military was sent out to prevent the human cargo from landing. In the standoff, the *Tampa* remained at sea while the government rushed through the legislation for a so-called Pacific Solution, diverting the asylum seekers to New Zealand and Nauru for processing their claims for refugee status.

24. Agence France-Presse, "Australia Switches Immigration Focus to Lure More Expert Workers," July 19, 2000.

25. *Tribute* (India), "Now New Zealand Wants IT Specialists," September 8, 2000.

26. *Wall Street Journal*, "Tug of War for IT Talent," April 11, 2001.

27. *Migration News*, "Singapore: IT Industry," 7, no. 5 (May 2000).

28. *Computer World*, "The New Immigration Waves," March 12, 2001.

29. *New York Times*, "Workers, and Bosses, in a Visa Maze," April 29, 2001.

30. *Bloomberg News (*New York), "Visa Process Frustrates Immigrants, U.S. Technology Companies," August 18, 2000.

31. The IT labor placement business in the developed world started in the late 1970s and early 1980s, initially driven by the sheer demand for IT expertise from the then nascent IT industry. (The placement business until then focused on semiskilled occupations such as secretaries and hotel assistants.) With the widening application of computerized systems in various commercial and industrial sectors in the 1980s, IT skills became unprecedentedly valuable, and as a result a new flexible employment pattern using "contract consultants" came into existence. Typically, contract consultants registered a one-person company under which they contracted out their own services to clients. Established placement agents in Australia at that time such as IDAPS saw business potential in this employment pattern and advocated it through seminars and the media, and, gradually, IT recruitment grew into a substantial business.

The employment pattern of contract consultants and the IT placement business were entrenched by the economic recession of 1990–92 when a great number of professionals throughout the capitalist world were thrown back into the labor market. Many from the ranks of traditional professions, such as mechanical engineering, switched to the growing computing sector, quite often starting out by taking on short-term contract assignments to gain basic hands-on experience. On the employer's side, being freed from the overhead of retaining permanent staff in the gloom of a recession made the hiring of contractors a business preference. By 2000, more than 30 percent of IT professionals in Australia worked on a contract basis (interview with Michael Bulter, statistics officer, Information Technology Professionals Association [ITPA], March 6, 2000, Melbourne) and, increasingly so, even senior managers (*Australian IT,* "Contracting Seen as Way of Entering IT Industry," June 30, 2000; citing a report by Touche Tohmatsu Deloitte 2000). By the turn of the century, up to 4,000 agents were involved in IT placement in Australia (*IT Jobs* [Australia]), "Headhunters Join Forces," November 6, 2001), and 50 percent of all IT companies relied on placement agents in labor management (interviews in Sydney with Terry Porter, senior manager, ICON Recruitment Pty. Ltd., May 11, 2000; and Stephen Leo, senior manager MBT Pty. Ltd., May 11, 2000; and in Melbourne with Michael Bulter, ITPA, March 6, 2000). Most IT job advertisements in the *Australian IT* (a section of the national newspaper the *Australian*) and the Web-based *Fairfax,* the two main IT job-information platforms in Australia, were posted by agents rather than IT companies. In the United States, at Microsoft's headquarters in Redmond, Washington, 6,000 of its 20,000 workers in 2000 were recruited and employed by placement agents rather than by Microsoft directly. Many workers had come to be known as "perma-temps" because they had worked for Microsoft for years but were still managed by agents (*Guardian,* "Workers' Right 'Abused in U.S.,'" August 30, 2000).

Chapter 2 Producing "IT People" in Andhra

1. Although Naidu (2000, 7–9) recalled that the meeting with Gates was finalized through the then U.S. ambassador to India, a few informants told me that Telugu IT professionals working in Microsoft were instrumental in arranging the meeting. The perception of the ubiquitous presence and global influence of Telugu IT people was prevalent in Andhra Pradesh.

2. *Newstime* (Hyderabad), "Babu among 'Seven Doing Wonders in the World,'" September 5, 2001.

3. *Deccan Chronicle* (Hyderabad), "Clarify State's Position in IT: Rasaiah," December 21, 1999.

4. *Computer Today,* "Indian Infotech Industry," July 1–15, 2001: 24.

5. NRI accounts were introduced by the Indian government in the 1970s, where NRI could deposit in foreign currencies with a high interest rate (for details, see Nayyar 1994, 51).

6. *Outlook India,* "The Best and the Brightest," May 29, 2000.

7. *Business Week,* "India's Whiz Kids," December 7, 1998.

8. The boom in IT education was greatly fueled by the state government helmed by Chandrababu Naidu, who had mandated that each of the twenty-three district capitals should have at least one engineering college—though, in effect, given the very limited government funds, these could only be private colleges (nominally

affiliated to a public university as required by government regulations). To stress the state's priority on IT education, Naidu went so far as to suggest that university degrees in the social sciences and humanities should be scrapped, which was roundly castigated as "uncivilized" by the Andhra Pradesh Teachers' Federation (*Deccan Chronicle,* "Naidu's Remarks Condemned," December 13, 1999) and led to agitations against him by staff from Osmania University and the University of Hyderabad (*Deccan Chronicle,* "Cyber CM Upset Academics," December 3, 1999).

9. Universities paid agents commissions for the recruitment, which for an Australian university was usually 10 percent of one year's tuition fee (AUD 1,000–1,500). On top of that, some agents charged students a fee of INR 500–5,000. Data from IDP (originally International Development Program, an umbrella association of universities in Australia devoted to promoting international recruitment) and my interviews with universities in Sydney suggest that some 6,000 students from India were enrolled in IT or IT-related courses in Australian universities in the year 2000. A survey by the International Office of the University of Western Sydney in 2000 found that as many as 80 percent of the Indian IT students enrolled had applied for admissions through agents; this was quite consistent with the survey of the University of Wollongong and Sydney University of Technology. In 1999, Interface, one of the largest India-Australia education agents, earned a revenue of AUD 28 million (*Sydney Morning Herald,* "Out of His Class," August 2, 2000).

10. Interviews with Mr. P. T. Prabhakar, deputy director, Commissionerate of Technical Education, July 31, 2001, Hyderabad; Mr. S. P. V. Sarabhaiah, officer on special duty, State Council of Higher Education, July 26 and August 6, 2001, Hyderabad; and Professor Subbar Reddy, Osmania University, July 26, 2001, Hyderabad.

11. Private college fees in Andhra Pradesh were charged in line with the Indian Supreme Court's verdict in the Unni Krishnan Case in 1992: 50 percent of seats were "free seats" allocated by the government according to examination results which charged INR 5,000 each as the annual tuition fee; 40 percents were paid seats (INR 35,000 per year) held by college management boards; and the remaining 10 percent were reserved for children of NRIs.

12. Osmania University is a public university that had 298 affiliated, private degree-granting colleges in 2000 (SHEC 2001c).

13. Most institutes adopted a "multibatch model," conducting courses daily for four or five batches with twenty to one hundred students each. (Interviews with G. Yoganand, managing director of Manjeera Estate Private Limited, the developer of Aditya Enclave, a residential complex now accommodating a large number of IT companies July 9, 2001; K. Krishnaiah, a shop owner and president of Aditya Enclave Residents' Association, July 7, 2001, Hyderabad.)

14. Unless otherwise indicated, information on NIIT was obtained through my interview with Anuj Rahul Joshi, regional manager NIIT Ltd., July 21, 2001, Hyderabad.

15. *Computer Today,* "Hot Skills, Cool Courses," May 1–15, 2001: 33.

16. Interviews with K. Krishnaiah of Aditya Enclave Residents' Association, Rajashekhar Reddy of I-Logic Institute, and L. A. Vandana, general manager, Trendz Information Technologies Ltd., an IT consultancy in Aditya Enclave, July 11, 2001, Hyderabad.

17. Some institutes sold course certificates in a last-ditch attempt to recoup their losses on closing down, like an institute reported as peddling one hundred certificates in the last ten days of August 2001. Certificates validated with the institute seal went for INR 200–2,000 and "blank" ones for INR 50–200 in 2001 (*Times of India, Hyderabad,* "Fake IT Certificates for Sale in City," August 31, 2001).

18. *Computer Today,* "Hot Skills, Cool Courses," May 1–15, 2001: 32 and 34.

19. Interview with Chandra Sheker, business executive, NIIT Tanuku. August 13, 2001, Tanuku.

20. A survey reported that 70 percent of the software professionals in big companies were from regions outside of metropolitan areas (*Dataquest,* "Reality? Not Anymore!" May 15, 2001: 65). Rajagopal (2000) discerned an association between the spread of the Vishwa Hindu Parishad (VHP) or World Hindu Council, and Hindu Swayamsevak Sangh (HSS) or Hindu Volunteers' Corps—both Hindu fundamentalist movements—in the United States on the one hand, and the influx of Indian IT professionals on the other, and attributed it to the fact that many IT workers were from small- and medium-sized towns that the author held to be more religious in outlook. VHP and HSS were also visible among the Indian IT professionals in Sydney. When Hong Kong was handed over to the People's Republic of China in 1997, the East and Southeast Asia headquarters of the VHP and HSS previously based there were relocated to Sydney. There were an estimated four hundred HSS members and eight hundred VHP members in Sydney in 2000, and as many as half of these activists could be IT people.

21. In contrast to the explosive expansion and dynamism of private IT education, the public primary-school education in Andhra Pradesh performed poorly. In West Godavari, the dropout rate at levels six to eight (ages eleven to thirteen) was as high as 70.64 percent in 1998 (DES 1999, table 19.7 [B]), and a news report revealed that even in the state capital only 20–60 percent of government school seats were filled in 2001, depending on the school (*Times of India, Hyderabad,* "Teachers in a Fix over Student Dropout Rate," September 3, 2001). Given the poor state of government primary education—the literacy rate in Andhra Pradesh was ranked twenty-seventh out of the twenty-eight states in India in 2001 (*Deccan Chronicle,* "AP Slides Down Literacy Scales," August 7, 2001)—it would have been miraculous if students had achieved the "merits" required by the global IT industry unless they'd been educated at private schools and colleges.

22. The central government of India categorizes the traditionally disadvantaged populations into three "backward" groups: Scheduled Castes (including those formerly considered "untouchable"), Scheduled Tribes, and Other Backward Classes. A reservation policy mandates that in public universities and government institutes, 15 percent of seats/jobs are reserved for Scheduled Castes, 7.5 percent for Scheduled Tribes, and 27 percent for Other Backward Classes.

23. The nine villages included five in West Godavari district—Undrajavaram, Doddipatla, East Vatpparrul Iragavaram, Peda Amiram, and Illindila Rapu; one in East Godavari—Atreyapuram; another in Prakasam district—Gurrapaadiya; and two in Mahaboob Nagar district in west Andhra Pradesh—Barhanpur and Tandra. As land is distributed unevenly, it is relatively easy to estimate the volume of the agricultural surplus of a village. For example in West Godavari, I was told

that 45–55 percent of the households had less than two acres of land, which generated little or no agricultural surplus. Only 5–10 percent of the households had more than five acres of land that could have generated a sizeable surplus. I also used the rate of land rent as a key index in quantifying surplus based on the assumption that the rent roughly equals the surplus (i.e., the tenants maintained a subsistence life after paying the rent), which was INR 10,000 per acre per annum on average in the coastal Andhra region. The limited number of persons who received tertiary education also made it possible to estimate the input in education. Besides higher education, the other two major sectors that absorbed agricultural surplus in the coastal area were the real estate and film industries.

24. Interviews with Shamar, Prabhakar, Sudarshan Rao, R. N. Reddy, July and August, Hyderabad (note 10, this chapter). Most private intermediate colleges in Andhra Pradesh were run by Kammas. Ownership over private degree colleges was more diverse, including Kammas, Reddys, and Rajus. In the case of Muslim-run private colleges, remittance from the Middle East and the West formed a major source of capital.

25. In Andhra Pradesh, many Brahmins lost their properties in the village as a result of the anti-Brahmin movement and had to move to cities, forming a significant stream of rural-urban migration.

26. My informants usually defined rural middle-class families as those holding three to six acres of land.

27. *Times of India, Hyderabad,* "Govt. Succeeds in Preventing Migration of Students to Neighbouring States," July 28, 2001.

28. Previously, the practice of bride price (the wealth paid by grooms to brides) was the norm for most castes in the south, except in part of Kerala (Srinivas 1983). Srinivas convincingly argued: "modern dowry is entirely the product of the forces let loose by British rule such as monetization, education and the introduction of the 'organized sector'" (1983, 13). (According to the definition adopted by official surveys in India today, the organized sector refers to all establishments in the public sector and the non-agricultural establishments in the private sector employing ten or more workers [see DES 1999, 413, table 28.17, footnote 1].) For example, soon after the introduction of Western education at the beginning of the twentieth century, a man with a bachelor of law degree from Calcutta University was already demanding a dowry of INR 10,000 (then GBP 700) (Srinivas 1983, 14). In short, dowry evolved in some sense as the price paid by wealthy families with daughters to purchase those heavily invested and highly profitable men in the "superior" organized sector, particularly the civil service. Thus, Srinivas (1967a, 126) saw the institution of dowry as part of the process of "secularization" (rather than "Sanskritization"), noting: "[a]nother evidence of increased secularization is the enormous importance assumed by the institution of dowry in the last few decades. Dowry is paid not only among Mysore and other South Indian Brahmins, but also among a number of high-caste groups all over India." Based on historical materials of the Punjab, Oldenburg (2003) demonstrated that modern dowry in northern India to a large extent owes its root to the exclusion of women from property rights over land, and the "masculinization" of the economy, which made men far more economically valuable than women—both were outcomes of British colonial rule.

29. I was told that in 2001, apart from an educated minority, the majority of the Mala and Madiga (two scheduled castes) population in Reddynagar district in northern Andhra Pradesh still paid bride price (*voli* in Telugu).

30. Exchange marriages, or *kunda maarpidi* (literally "pot exchange" in Telugu), are matches between one pair of siblings and another.

31. *Newstime*, "UNICEF Sees Many Grey Areas in A.P.'s Social Development," August 1, 2001.

32. Padmashali is a poor weaving caste in Andhra Pradesh. They were frequently in the news in the 1990s when many committed suicide because their traditional hand-loom livelihoods were destroyed by the introduction of power looms in the state.

33. The husband could demand dowry at any time as long as the marriage was alive. The opportune moment to raise demands came when married women visited their natal family on festival occasions such as *rakshabandhan* (brother-sister festival in early August, previously a northern Indian custom but now increasingly popular in the south) or *grihapraresh* (house-entering ceremony mainly performed by sisters when the brothers move into a new house).

34. High dowries for IT grooms were not confined to the forward castes. Owing to the reservation policy, members of the backward castes have also moved up to become professionals, including IT people. In order to confirm their newly achieved status, these families were sometimes even more anxious than, say, the Kammas to pay high dowries. I was told of a well-known lawyer in Hyderabad from the Mala community who offered an astonishingly high dowry for his daughter to marry into a Kamma family. Additionally, the dowry custom was also practiced among the Muslim community in Hyderabad, which constituted roughly 39 percent of the city's population of 3.1 million in 2000 (see http://www.andhrapradesh.com, table 1.21). The rates were normally lower than those of the Hindus: the dowry for a Muslim IT groom in 2000 ranged from INR 30,000 to INR 1 lakh. Muslim weddings, however, were lavish affairs that normally invited contributions of INR 2–3 lakhs from the bride's family and about INR 1 lakh from the groom's. Similar to the significance of dowry for Hindus, these expenditures served as an indicator of the family's social status, and therefore consumed a large part of the remittance from sons working in the Middle East or the West.

35. *Computer Today*, "In a Nasty Through," October 2001: 105.

Chapter 3 Selling "Bodies" and Selling Jobs

1. Although I emphatically identified myself as a Chinese from China studying in Oxford, my Indian informants were far more interested in my being "from England"—a provenance that also was glossed as "from London" by some who tried to help me attract more cooperative informants.

2. Laxman, a thirty-nine-year-old software engineer working in the Australian branch of Novell, a global provider of net-services software headquartered in San Jose, California, had in his first year in Sydney lived in Liverpool, a large southwest suburb where many of the Indians were mechanical engineers. Disclosing his

Brahmin background, he began by choosing his words carefully when describing to me the general situation of the Indian professionals in Sydney:

> There are two . . . kinds of Indians here. IT people and other conventional technicians are different. People in Liverpool are stuck. They can't go back. If *we* go back, life is similar. But the Liverpool people can't even maintain a two-wheeler there. From India to Australia is a big jump for them. . . . We don't mix too much. *They* stick together—who found a new job, who took an IT course—gossip all the time. They can't see beyond the corner where they live. *Real IT people* have their professional line. We are really looking at the world. . . . You can say these are two classes!

3. A "walk-in interview" job fair held in Bangalore in mid-2001 attracted about 1,000 IT hopefuls, who paid for an entrance ticket of INR 200, only to find out that the event was a cash-collecting scam—there were no legitimate IT firms signing up workers (cited in Desilva 2001).

4. *CNET News.com* (news.com.com), "H-1B Visas Jump in 2001," January 22, 2002.

5. *Times of India, New Delhi,* "Destination United States May Not Be Easy Any More," October 5, 2001.

6. See for example, *San Jose Mercury News,* "Back to India for Tech Worker," February 3, 2002.

7. The State Department of the United States assessed one fifth of the H-1B applications received in India in 1998–99 to contain false information. An appeal for help from Indian software firms and trade associations by the consulate in Chennai resulted in an "Online Information Verification Engine" (codenamed "Project Olive") containing data on all legitimate graduates from colleges in Hyderabad, as fake documents were particularly endemic in the city (see *San Jose Mercury News,* "Predatory Culture of Fraud, Abuse Emerging in India," November 19, 2000.).

8. The price paid for a job in Hyderabad was much higher than that in Bangalore, see *Times of India, Hyderabad,* "Jobs Aplenty . . . if Wallets Are Loaded," July 12, 2001.

9. SAP is a global software company headquartered in Germany specializing in business solutions; SAP courses provide the professional qualifications for running these programs.

Chapter 4 Business of "Branded Labor" in Sydney

1. The first Indian immigrants to Australia were brought in by British colonists during the period 1800–1901, and these arrivals ceased in 1901 after the adoption of the "White Australia" policy, as the Immigration Restriction Act came to be known. A second wave, beginning with the movement of Anglo-Indians (offspring of Indians and Europeans) in the 1950s in the wake of Indian Independence and enlarging as the "White Australia" policy was relaxed in the 1960s, comprised

large numbers of medical doctors. The subsequent wave, particularly from the 1980s, was mainly composed of skilled immigrants, including engineers, IT people, and those who were recruited through body shopping after the mid-1990s.

2. As my informants estimated, in Sydney in 2000, less than 5 percent of Indian IT professionals counted among the most senior 20 percent positions in their organizations, and about 80 percent of the professionals held middle-rank positions. Surveys by *Dataquest* in March 1989 (cited in Lakha 1991, 32) and Lakha (1991) came to similar estimates.

3. I was told that ethnic-network-based subcontracting of this kind was prevalent in the United States as well. Spouses of H-1B visa holders (who held H-4 visas that precluded employment) often had IT skills, learned in India or from their husbands, and formed an important subcontracting labor force for small Indian IT businesses.

4. There were quite a few home-based training courses in Sydney; most of them were run as a part-time business, and equipped with one or two computers, adequate for up to three students at a time. Home-based instructors sometimes associated with one another to offer a wider range of courses. For example, Kaaveri, a dynamic female IT professional running a course at home, introduced students wanting courses that she could not deliver to one of her six associated teachers; she collected the fees from the students, deducted 25 percent and passed the rest on to the associate. Together, body shops and home-based classes in Sydney produced approximately four hundred to five hundred Indian IT workers in the year 2000–2001.

5. The documentation of medical treatments and consultations, a time-consuming task, has been increasingly outsourced to India by hospitals in the West. The typical practice is for a doctor in the United States to record an oral description or prescription, send the file through high-speed data line to India, where "medical transcriptionists" (MTs—an established term in urban India by 2000) transcribe the dictation and send it back as a document file. Medical-records transcription in India was an industry in itself with a record revenue of USD 38 million in 2002 (Nasscom, cited in Press Trust of India, "Healthcare Outsourcing on High despite BPO Backlash," June 29, 2004). Many media reports highlighted the twelve-hour time difference between India and the East Coast of the United States that supposedly enabled the business to run around the clock without creating any delay in the work schedule in the United States. This is more a myth than truth. Transcription is a very tedious job and one record usually needs to be checked three times and sometimes takes up to a week to complete. Due to the competition among the numerous transcription companies in India, the charge for the service dropped continuously, to USD 0.05 for each line transcribed in 2001, and MTs were paid around INR 5,000 a month (interviews with Vishnu, July 24, July 26, and August 30, 2001, Hyderabad).

6. It was quite common for technopreneurs from India to tour Western markets for business opportunities, hawking résumés of IT workers in India, and seeking contracts for outsourcing from large firms. One estimate suggested that at any given time in 1999, about 2,000 Indian technopreneurs were touring the United States for this purpose (Rajagopal 2000, 482).

Chapter 5 Agent Chains and Benching

1. Placement agents normally divide their staff into two groups: account managers responsible for maintaining smooth relationships with clients, and recruitment consultants in charge of recruiting workers who take direction from their account managers regarding their client's needs.

2. Interviews with managers of three major IT placement agents in Australia: Drake International Australia Pty. (October 10, 2000), ICON Recruitment Pty. Ltd. (June 16, 2000), and Morgan and Banks Technology Pty. (March 3, 2000), all in Sydney.

3. Interviews with a manager of Mastech Australia, August 1, 2000, and February 4, 2001, Sydney.

4. Manpower, set up in the United States in 1948, was world's largest provider of workforce management and services in 2000, with over 3,700 offices in fifty-nine countries (*IT People Evolve,* "IT Opportunities," 2001, 3[49]: 9).

5. *ITemploymentNEWs* (Australia), "Why the IT Job Agents Have a Shrinking Feeling," June 20, 2000.

6. Spotter's fee in 2000 and early 2001 ranged from AUD 200 to AUD 500 in Australia for each worker introduced, but could be as high as USD 5,000 in the United States around that time. Ridgeway, a U.K.–based multimedia company, offered GBP 1,000 to any member who could suggest a suitable IT employee in 2000 (BBC, "Wanted: Somebody . . . Anybody? Remember Unemployment? Now Many Jobs Go Unfilled," September 11, 2000).

7. The Sunday morning religious lectures (in Tamil or English) in Sydney had been in place since the early 1990s. In 2001, it drew twelve to fifteen adults together with a separate class for five groups of children, organized according to age. The series was run by a group of devoted individuals and was closely associated with the Westmead Temple in Sydney run by Sri Lankan Tamils.

8. One ruse was subletting accommodations to workers and declaring the rents paid for the workers as part of tax-deductible business expenditures, though in reality, the body shop collected rents from workers, sometimes even higher than what they paid!

9. *San Francisco Chronicle,* "New H-1B Visa Law a Life Changer," October 5, 2000.

Chapter 6 Compliant Bodies?

1. According to the conditions of H-1B visas, employment-based green cards are issued to those who have worked in the United States with the same sponsor for six years. Thus, a change of job/sponsor while a worker's green-card application is pending would automatically cancel the time accumulated under the previous sponsor, and the worker would have to wait another six years to be eligible for a green card.

2. "Gross salary" here refers to the amount received by the last agent in the chain (i.e., the sponsor), and this was also what "salary" meant in most contracts

between workers and body shops. Workers were usually made to believe that the substantially reduced "net salary" amount they finally received was due to the deductions of nearly 50 percent for taxes and 10 percent as the body shop's profit. In reality, about 30 percent of the workers' gross salary went for taxes, and another 30 percent was extracted by the body shop sponsor. As evidence, body shops often cited the wages of Australian permanent workers, which were not much higher than the net salary given by body shops. In fact, 457 visa holders could claim a substantially lower tax rate under a "living away from home allowance"; furthermore, a temporary contract worker was paid more than double the monthly wage of a permanent worker of the same level because the former had no fringe benefits. In addition, I did not come across any body shop that paid superannuations or insurance—another often claimed deduction—for its workers.

3. The position taken by the body-shopping middlepersons stood in sharp contrast to my experience researching a migrant community in Beijing (Xiang 2000). In the latter context, I had seen how middlepersons tend to have lasting influence over the relation between the worker and the boss; if the worker was treated badly by the boss, the middleperson would either pressure the boss, or, if possible, do something to help both the boss and worker for example by helping the boss get more business and then the worker get better pay. If the middleperson failed to change the situation, the middle person would feel that he or she had lost "face": the boss had not cared properly for the worker because the boss had not taken the middleperson seriously!

4. In Andhra Pradesh, as I was told by Uday, when people achieved something they could not hide, such as a new house, or when they wanted to display achievements to claim status, they held special *puja* to avert others' "*asuya*" or "*disty*"—special terms in Telugu for envy with harmful effects.

5. "Exit" refers to the strategy of employees or customers to vote with their feet; "voice" is when members of an organization bring up dissenting concerns within the organization without leaving it, which is a political action par excellence. From a collective-action point of view, exit is likely to be more efficient than voice: dissatisfied customers in most cases get their message across better by boycotting products than by voicing their concerns with the producer. But exit is not always a practical solution for lapses in a system, particularly when it involves employer and employees. Hence, "loyalty": employees remind their employers of what they are unhappy about and employers keen to encourage loyalty take heed. Therefore, lapses can be corrected at the right time, and the system as a whole be sustained.

Chapter 7 The World System of Body Shopping

1. Dasara in October is an important event in Mysore, Karnataka, southern India, often lasting for one month. The festival celebrates the victory of good over evil, marking the slaying of the demon Mahishasura by the goddess Chamundeshwari.

2. Much earlier, David Hale, the chief global economist with Zurich Financial Services in Chicago, had assessed the vulnerability of the Australian dollar as being

due to the weakness of the country's IT sector, and recommended increases in skilled immigration (see Birrell 2000, 78).

3. Business colleges, which have mushroomed in Australia since the 1980s, are private colleges offering one- to two-year business-diploma courses, and after 1998, many adopted IT as their core program. Business colleges charged much lower tuitions than universities, but demanded high fees for admission. Recruitment agents in Hyderabad, many set up by returning college students, were given INR 10,000–20,000 for each student recruited in 2001. Some colleges sent their "recruitment teams" to the Sydney airport, approached anyone who looked like an Asian student, promised him or her a tuition fee lower than others, and poached them straightaway. Due to the low educational standard and reputation, business-college graduates were unlikely to become PR on graduation, and applying for 457 visas was a more feasible means for staying on; they thus constituted important onshore clients of body shops in Australia.

4. Ashok used this phrase to describe the value of migrating to the United States when we were discussing his rationale for purchasing a house in Strathfield in Sydney together with Uday: the house was costly but had a high resale value, which was most important, he said, as they could be moving to the United States "at any time."

5. The case of H-4 visa holders who face immediate deportation in the case of divorce has left many Indian H-4 wives, the majority married to IT husbands, trapped in abusive marriages. The San Francisco-based South Asian women's organization, Maitri, received over 1,500 calls from H-4 spouses reporting domestic violence over the period of 2000–2001 (*Times of India, Hyderabad,* "H4 Visa Leaves Many Indian Women Dependent on Abusive Husbands," July 22, 2001).

6. *Asian Wall Street Journal,* "Work Policy Worries Some Singaporeans," October 9, 2001.

7. Satyam established its Asia Pacific headquarters in Singapore in 2001. In the same year the Singapore government inaugurated the India Center to assist Indian enterprises, especially IT companies, to set up operations in Singapore.

8. This multidirectional mobility was dramatically intensified during the 2001 market slowdown. Sumanth, a Telugu student in Sydney, asked Uday to help forward his brother's résumé to body shops in Australia when the brother, an IT professional, lost job in Singapore in June 2001. Two weeks later, the brother went to Kuala Lumpur to search for jobs himself. When I tried to contact him before I myself headed for Kuala Lumpur, he had already gone on to Bangkok!

9. Palm Court was apparently so well known that the condominium was raided by Malaysian police in March 2003, and 270 Indian IT professionals were arrested for alleged visa irregularities. This created a diplomatic standoff between Malaysia and India and prompted then Singaporean prime minister Goh Chok Tong to announce in an interview with the newspaper the *Hindu* that "Indian IT professionals are welcome here [in Singapore]. Please keep it on the record" (Agence France-Presse, "S'pore Welcomes Indian IT Experts," April 7, 2003).

10. All my informants talked about income in the Middle East in terms of Indian rupees instead of the local dinars or riyals, but used the local currency or American dollar when discussing Australia, Singapore, or Canada.

11. In Saudi Arabia, even those working for multinational companies needed a Saudi citizen in a senior position in the company to stand as a personal guarantor, and a "no-objection certificate" from the guarantor was a precondition for changing jobs. If a request to leave a company was rejected, no further requests would be entertained for the next eighteen months, necessitating those who wanted to move to another country in the same region to return to India to apply for a fresh work visa.

12. In 2000 alone, the Jamaican government invested USD 280 million to support three U.S. IT firms to set up operations there (*Jamaica Gleaner,* "Editorial, Information Technology Jobs," August 20, 2000). The U.S. Embassy and the U.S. Agency for International Development (USAID) also played active roles in facilitating nearshore development (*Jamaica Observer,* "$6m Training Project for Local Businesses," November 29, 2001).

13. *Jamaica Observer,* "Gov't to Train More Computer Programmers," November 21, 2001.

14. "Are You Indian?" was the line printed on the T-shirts of the staff from the Berlin-based Internet firm Datango who were in Bangalore in 2000 to recruit workers; the back of the T-shirt carried the URL for the company's job listing. See *Time Europe,* "Germany's New Recruits," June 25, 2001.

15. Center for Immigration Studies News Services (Washington), "German Cabinet Approves Plan to Allow More IT Workers," May 31, 2000. The new program was dubbed the "German green-card scheme" in the media and in official statements, though it in no way resembled the U.S. scheme, which prompted Venu to suggest it was a devious "German tactic" to lure Indian workers by confusing them into thinking otherwise!

16. *Frankfurter Allgemeine Zeitung,* "Green Card Holders Settle into Germany," July 25, 2001.

17. *Time Europe,* "Germany's New Recruits," June 25, 2001.

18. *EE Times,* "U.S. Finds Itself Competing for Indian Engineers," June 30, 2000.

19. *Times of India, Hyderabad,* "Indians Get Largest Number of German Green Cards," August 1, 2001.

20. *Time Europe,* "Germany's New Recruits," June 25, 2001.

21. *Financial Times,* "Tokyo to Lure Foreign IT Skills," May 31, 2001.

22. *Times of India,* "Korea Beckons IT Professionals," August 3, 2001.

23. *San Jose Mercury News,* "Can There Be a German Dream? Laid-Off H-1B Workers Face Making Yet Another New Home in a Foreign Country," August 19, 2001.

24. Agence France-Presse, "Indian Software Experts Look Askance at Germany's Welcome Mat," May 26, 2000.

Ending Remarks The "Indian Triangle" in the Global IT Industry

1. In California for example there were more than 7,000 high-tech companies run by Indians in 2000 alone, generating an estimated USD 60 billion in sales (Reuters, "India's IT Industry Geared to Move Up Value Chain," May 9, 2001).

2. Many families that I came across in Andhra Pradesh still did not allow so-called low-caste maids to cook or even enter the kitchen. Sometimes one family hired more than one maid: the so-called lower-caste one for cleaning and those from castes compatible to the employers' for cooking. One friend of mine cited it as a proof for the Hyderabadi cosmopolitan progressiveness that maids there were allowed to walk through the same front door as their masters and to sit aside the master occasionally.

3. India has the largest child-labor population in the world, and Andhra Pradesh has the largest in India. One quarter of the children aged five to fourteen in Andhra Pradesh fell within the global definition of child labor in 2001 (UNICEF, cited in *Newtimes,* "UNICEF Sees Many Grey Areas in A.P.'s Social Development," August 1, 2001), and the International Labour Organization estimated the population of child labor in the state to be 1.66 million at around the same time (cited in *Deccan Chronicle,* "AP Slides Down Literacy Scales," August 7, 2001).

4. The so-called low-caste and low-class families were often more vulnerable to the negative consequences of globalization and IT industry than those "in" the New Economy. For example, the Lambada tribe in the north of the Telangana region of Andhra Pradesh produced a few professionals including IT workers. Instead of helping the community prosper, this, compounded by the tribe's pursuit of converting themselves into the Reddy caste, made dowry a new custom of the community. Despite the poverty, a dowry of INR 40,000–50,000 was common and sometimes went up to INR 1 lakh. Women had to earn money to marry themselves off, and the tribe had thus allegedly become a source of sex workers in Andhra Pradesh. In some extreme cases, I was told, the husband took the dowry and subsequently abandoned the wife with the children on the grounds that she had been a prostitute. The man may have then moved on to marry another one with another dowry.

Alston, R. 1999a. "E-commerce: The Banking Business Partnership." Speech at the E-commerce: The Banking Business Partnership Forum. Sydney. October 22.
———. 1999b. "The Role of Communications and E-commerce in Building the Nation." Speech at Minter Ellison Nation Builders Awards Ceremony. Melbourne. March 25.
Alston, R., D. Kemp, P. Reith, and P. Ruddock. 1999. Joint press release: "Providing for the Future IT&T Skill Needs of Australia." April 29.
———. 2001. Joint press release: "Immigration Initiatives to Attract IT Experts." January 29.
Appelbaum, E., and C. Rouse. 2001. "Easing Restrictions on Visas Doesn't Help Any High-Tech Workers." In *EPI Viewpoint*. Washington, D.C: Economic Policy Institute. Available at http://www.epi.org/content.cfm/webfeatures_viewpoints _h1-bvisas.
Atkinson, R., and R. Court. 1998. *The New Economy Index*. Washington, D.C.: Progressive Policy Institute.
Barrett, R. E., and S. Chin. 1987. "Export-Oriented Industrializing States in the Capitalist World System: Similarities and Differences." In *The Political Economy of the New Asian Industrialism*, ed. F. Deyo: 23–43. Ithaca, N.Y.: Cornell University Press.
Bauman, Z. 1998. *Globalization: The Human Consequences*. New York: Columbia University Press.
Bhachu, P. 1988. "Apni Marzi Kardhi. Home and Work: Sikh Women in Britain." In *Enterprising Women: Ethnicity, Economy, and Gender Relations*, eds. S. Westwood and P. Bhachu: 76–102. London and New York: Routledge.
Birrell, B. 1999. "The 1999–2000 Immigration Program." *People and Place* 7(2): 52–53.

———. 2000. "Information Technology and Australia's Immigration Program: Is Australia Doing Enough?" *People and Place* 8(2): 77–83.

Boyd, M. 1989. "Family and Personal Networks in International Migration: Recent Developments and New Agendas." *International Migration Review* 23:638–70.

Brettell, C. 2000. "Theorizing Migration in Anthropology." In *Migration Theory,* eds. C. Brettell and J. Hollifield: 97–135. New York: Routledge.

Cairncross, F. 1997. *The Death of Distance: How the Communications Revolution Will Change Our Lives.* London: Orion Business Books.

Carrier, J. 1998a. "Introduction." In *Virtualism,* eds. J. Carrier and D. Miller: 1–24. Oxford: Berg.

———. 1998b. "Abstraction in Western Economic Practice." In *Virtualism,* eds. J. Carrier and D. Miller: 25–48. Oxford: Berg.

Carrier, J., and D. Miller, eds. 1998. *Virtualism.* Oxford: Berg.

Castells, M. 1996. *The Rise of the Network Society.* New York: Blackwell Press.

———. 2001. *The Internet Galaxy: Reflections on the Internet, Business, and Society.* New York: Oxford University Press.

Castles, S. 2001. "Studying Social Transformation." *International Political Science Review* 22(1): 13–32.

Clarke, J., and J. Salt. 2003. "Work Permits and Foreign Labour in the UK: A Statistical Review." *Labour Market Trends* (UK Home Office), November: 563–74.

Cohen, Y., and Y. Haberfeld. 1993. "Temporary Help Service Workers: Employment Characteristics and Wage Determination." *Industrial Relations* 32(2): 272–87.

CTE (Commissionerate of Technical Education), Andhra Pradesh. 2001. "Increase in IT Training Institutes by Year." Unpublished document.

DCITA (Department of Communications, Information Technology and the Arts), Department of Education, Training and Youth Affairs, Department of Employment, Workplace Relations and Small Business and Department of Immigration and Multicultural Affairs, Australia. 1998. "Skill Shortages in Australia's IT&T Industries." Canberra: Discussion paper.

Department of Industry, Science and Tourism, Australia. 1998. *Information Industries Action Agenda.* Canberra: Commonwealth of Australia.

DES (Directorate of Economics and Statistics), Andhra Pradesh. 1999. *Statistical Abstract of Andhra Pradesh, 1999.* Hyderabad: Directorate of Economics and Statistics.

Desilva, B. (representative of International Telecommunication Union). 2001. Speech at the Andhra Pradesh Information Technology Professionals' Forum. Vyceroy Hotel, Hyderabad. July 25.

DIMA (Department of Immigration and Multicultural Affairs), Australia. 2000. "Immigration Program."

Eischen, K. 2000. "Information Technology: History, Practice and Implications for Development." Working paper of Center for Global, International and Regional Studies WP #2000–4. University of California, Santa Cruz.

Eriksen, T. ed. 2003. *Globalisation: Studies in Anthropology.* London and Sterling, Va.: Pluto Press.

Fields, G. 1984. "Employment, Income Distribution and Economic Growth in Seven Small Open Economies." *Economic Journal* 94:74–83.

Finance and Planning Department, Andhra Pradesh. 1999. *Information Technology Policy of the Government of Andhra Pradesh.* Hyderabad: Finance and Planning Department of Andhra Pradesh, May 25.

Gottfried, H. 1991. "Mechanisms of Control in the Temporary Help Service Industry." *Sociological Forum* 6(4): 699–713.

Granovetter, M. 1985. "Economic Action and Social Structure: The Problem of Embeddedness." *American Journal of Sociology* 91:481–510.

Heeks, R. 1996. *India's Software Industry.* New Delhi: Sage Publications.

———. 1998. "The Uneven Profile of Indian Software Exports." Development Informatics working paper series, no. 3. Institute for Development Policy and Management, University of Manchester, UK.

Herman, H. V. 1979. "Dishwashers and Proprietors: Macedonians in Toronto's Restaurant Trade." In *Ethnicity at Work,* ed. S. Wallman: 71–92. London and Basingstoke: Macmillan.

Hirschman, A. 1970. *Exit, Voice, and Loyalty.* Cambridge, Mass.: Harvard University Press.

Hirst, P., and G. Thompson. 1996. *Globalisation in Question.* Cambridge: Polity Press.

Hoogvelt, A. 1997. *Globalization and the Postcolonial World.* Baltimore, Md.: Johns Hopkins University Press.

INS (Immigration and Naturalization Services), United States. 2000. "Characteristics of Specialty Occupation Workers (H-1B): May 1998 to July 1999." Washington: U.S. Immigration and Naturalization Services.

Iredale, R. 1998. *Skill Transfer.* Wollongong: University of Wollongong Press.

IT&T Skills Taskforce, Australia. 1999. *Future Demand for IT&T Skills in Australia. 1999–2004.* Canberra: IT&T Skills Taskforce.

ITAA (Information Technology Association of America). 2001. Press release: "Major New Study Documents Demand for IT Workers Continues." April 2.

Kalleberg, A. L. 2000. "Nonstandard Employment Relations: Part-Time, Temporary and Contract Work." *Annual Review of Sociology* 26:341–65.

Kloosterman, R., and J. Rath. 2001. "Immigrant Entrepreneurs in Advanced Economies: Mixed Embeddedness Further Explored." *Journal of Ethnic and Migration Studies* 27(2): 189–202.

Kosmin, B. 1979. "Exclusion and Opportunity: Tradition of Work amongst British Jews." In *Ethnicity at Work,* ed. S. Wallman: 37–70. London: Macmillan.

Krishna, V. V., and B. Khadria. 1997. "Phasing Scientific Migration in the Context of Brain Gain and Brain Drain in India." *Science, Technology & Society* 2(2): 348–85.

Krissman, F. 2005. "*Sin Coyote Ni Patrn:* Why the 'Migrant Network' Fails to Explain International Migration." *International Migration Review* 39(1): 4–44.

Lakha, S. 1991. "Indian Computer Professionals in Australia." *BIR Bulletin* 5: 32–33.

Leyton, E. 1970. "Composite Descent Groups in Canada." In *Readings in Kinship in Urban Society,* ed. C. C. Harris: 179–86. Oxford: Pergamon.

Martin, P. 1996. "The Death of Geography." *Financial Times.* February 22.

Massey, D. 1990. "Social Structure, Household Strategies, and the Cumulative Causation of Migration." *Population Index* 56(1): 3–26.

———. 1994. "An Evaluation of International Migration Theory: The North American Case." *Population and Development Review* 20(4): 699–751.

Massey, D. S., R. Alarcon, J. Durand, and H. Gonzalez. 1987. *Return to Azatlán: The Social Process of International Migration from Western Mexico*. Berkeley: University of California Press.

Massey, D. S., J. Arango, G. Hugo, A. Kouaouci, A. Pellegrino, and J. E. Taylor. 1993. "Theories of International Migration: A Review and Appraisal." *Population and Development Review* 19(3): 431–66.

Massey, D. S., L. Goldring, and J. Durand. 1994. "Continuities in Transnational Migration: An Analysis of Nineteen Mexican Communities." *American Journal of Sociology* 99:1492–533.

Matloff, N. 1998. *Debunking the Myth of a Desperate Software Labor Shortage: Testimony to the U.S. House Judiciary Committee Subcommittee on Immigration*. Presented on April 21. Available at http://heather.cs.ucdavis.edu/itaa.real.html.

Meredith, D., and B. Dyster. 1999. *Australia in the Global Economy*. Cambridge: Cambridge University Press.

Ministry of Foreign Affairs, Japan. 2001. "Expansion of Issuing Visas for Business People of IT-Related Enterprises in India." February 2. Available at http://www.mofa.go.jp/announce/announce/2001/2/0202.html.

Moretti, E. 1999. "Social Networks and Migrations: Italy 1876–1913." *International Migration Review* 33(3): 640–57.

Naidu, N. C. (with Sevanti Ninan). 2000. *Plain Speaking*. New Delhi: Viking.

Nasscom (National Association of Software and Service Companies), India. 2000. "IT Software & Service Industry in India Grows by 53% in 1999–2000." Available at www.nasscom.org.

Nayyar, D. 1994. *Migration, Remittances and Capital Flow: The Indian Experience*. New Delhi: Oxford University Press.

NIIT (National Institute of Information Technology), India. 2001. "NIIT Facts." Unpublished document provided by NIIT Hyderabad, July.

Nolan, P. 2001. "The Changing Nature of Employment." Working paper of Australian Centre for Industrial Relations Research and Training, Sydney.

OECD. 2001. *Trends in International Migration 2000* (SOPEMI 2000). Paris: OECD.

———. 2002. *Trends in International Migration 2001* (SOPEMI 2001). Paris: OECD.

Oldenburg, V. T. 2003. *Dowry Murder: The Imperial Origins of a Cultural Crime*. New Delhi: Oxford University Press.

Pinkstone, B. 1992. *Global Connections: A History of Exports and the Australian Economy*. Canberra: Australian Government Publishing Service.

Polanyi, K. 1944. *The Great Transformation: The Political and Economic Origins of Our Time*. Boston: Beacon.

———. 1957a. "Aristotle Discovers the Economy." In *Trade and Market in the Early Empires: Economies in History and Theory*, eds. K. Polanyi, C. M. Arensberg, and H. W. Pearson: 64–94. Glencoe, Il.: The Free Press.

————. 1957b. "The Economy as Instituted Process." In *Trade and Market in the Early Empires: Economies in History and Theory,* eds. K. Polanyi, C. M. Arensberg, and H. W. Pearson: 243–70. Glencoe, II.: The Free Press.

Population Division, Department of Economic and Social Affairs at the United Nations. 2003. *Trends in Total Migrant Stock: The 2003 Revision.* New York: United Nations.

Poros, M. 2001. "The Role of Migrant Networks in Linking Local Labour Markets: The Case of Asian Indian Migration to New York and London." *Global Networks* 1(3): 243–59.

Portes, A., ed. 1995. *The Economic Sociology of Immigration: Essays on Networks, Ethnicity, and Entrepreneurship.* New York: Russell Sage Foundation.

————. 1999. "Conclusion: Towards a New World—The Origins and Effects of Transnational Activities." *Ethnic and Racial Studies* 22(2): 463–77.

Portes, A., and J. Sensenbrenner. 1993. "Embeddedness and Immigration: Notes on the Social Determinants of Economic Action." *American Sociological Review* 98(6): 1320–50.

Portes, A., E. Guarnizo, and P. Landolt. 1999. "The Study of Transnationalism: Pitfalls and Promise of an Emergent Research." *Ethnic and Racial Studies* 22(2): 217–38.

Pries, L. 2001. "The Approach of Transnational Social Spaces: Responding to New Configurations of the Social and the Spatial." In *New Transnational Social Spaces:* 3–33. London: Routledge.

Rajagopal, A. 2000. "Hindu Nationalism in the U.S.: Changing Configurations of Political Practice." *Ethnic and Racial Studies* 23(3): 467–96.

Raju, M. R. 1999. "High-LET Radiotherapy to Rural Health Care." *Journal of Radiation Research.* Vol. 40, Spring: 74–84.

Robison, W. I., and J. Harris. 2000. "Towards a Global Ruling Class? Globalisation and the Transnational Capitalist Class." *Science & Society* 64(1): 11–54.

Ruddock, P. 1999. "The Coalition Government's Position on Immigration and Population Policy." *People and Place* 7(4): 6–16.

Saxenian, A. 2001. "Silicon Valley's New Immigrant Entrepreneurs." In *The International Migration of the Highly Skilled,* eds. W. Cornelius, T. Espenshade, and I. Salehyan: 197–233. La Jolla: Centre for Comparative Immigration Studies, University of California–San Diego, San Diego.

SCHE (State Council of Higher Education), Andhra Pradesh. 2001a. "Growth of Engineering Colleges in Andhra Pradesh (as on 4–12–2000)." Unpublished document.

————. 2001b. "List of Engineering Colleges Established up to 1990, during the Period 1990–1995 and 1995–2000 in the State," Unpublished document.

————. 2001c. "Total No. of Degree Colleges in Andhra Pradesh during the Years 1990, 1995 and 2000." Unpublished document.

————. 2001d. "Number of Institutions and Intake in Undergraduate, Postgraduate and Professional Colleges in 13 Universities in Andhra Pradesh." Unpublished document.

————. 2001e. "Education Institutes Offering IT-Related Courses in AP." Unpublished document.

Simmons, A., and D. E. Plaza. 1999. "Breaking through the Glass Ceiling: The Pursuit of University Training among Afro-Caribbean Migrants and Their Children in Toronto." *Canadian Journal of Ethnic Studies* XXX(3): 99–120.

Sklair, L. 1995. *Sociology of Global System.* London and Baltimore: Prentice Hall and John Hopkins University Press.

———. 2001. *The Transnational Capitalist Class.* Oxford: Blackwell.

Software Technology Parks of India, Hyderabad. 2000. "Foreign Investment in the IT Sector, Andhra Pradesh," Unpublished document.

Srinivas, M. 1967a. "Secularization." In *Social Change in Modern India:* 118–46. Berkeley: University of California Press.

———. 1967b. "Westernization." In *Social Change in Modern India:* 46–88. Berkeley: University of California Press.

———. 1983. *Some Reflections on Dowry.* New Delhi: Oxford University Press published for the Centre for Women's Development Studies.

———. 1996. "The Quality of Social Relations." In *Indian Society through Personal Writings:* 163–91. New Delhi: Oxford University Press.

Tilly, C. 1990. "Transplanted Networks." In *Immigration Reconsidered: History, Sociology and Politics,* ed. V. Yans-McLaughlin: 79–95. New York, Oxford: Oxford University Press.

United Nations International Economic and Social Affairs. 1992. *International Migration Policies and Programmes.* New York: United Nations.

Vajpayee, A. B. 1998. Speech at the inauguration of Satyamonline. Hyderabad, November 24.

Vijayabaskar M., S. Rothboech, and V. Gayathri. 2001. "Labour in the New Economy: Case of the Indian Software Industry." *The Indian Journal of Labour Economics* 44(1): 39–54.

Waldinger, R., H. Aldrich, and R. Ward. 1990. "Opportunities, Group Characteristics, and Strategies." In *Ethnic Entrepreneurs:* 13–48. Newbury Park, Calif.: Sage.

Wallman, S. 1975. "Kinship, A-Kinship, Anti-Kinship, Variation in the Logic of Kinship Situations." *Journal of Human Evolution* 4:331–41.

Watson, I. 1996. *Opening Glass Doors: Overseas Born Managers in Australia.* Canberra: Bureau of Immigration and Population Research.

Werbner, P. 1988. "Taking and Giving: Working Women and Female Bonds in a Pakistani Immigrant Neighbourhood." In *Enterprising Women: Ethnicity, Economy, and Gender Relations,* eds. S. Westwood and P. Bhachu: 177–202. London and New York: Routledge.

White House. 1997. "A Framework for Global Electronic Commerce." July 1. Available at http://www.technology.gov/digeconomy/framework.htm.

Xiang Biao. 2000. *Kuayue Bianjie de Shequ: Beijing "Zhejiangcun" de Shenghuo Shi* (Transcending Boundaries: The Life History of "Zhejiang Village" in Beijing). Beijing: *Sanlian Shudian.*

Zahniser, S. S. 1999. *Mexican Migration to the United States: The Role of Migration Networks and Human Capital Accumulation.* New York and London: Garland Publishing.

FORMATION *Series*

Everything Was Forever, Until It Was No More:
The Last Soviet Generation
BY ALEXEI YURCHAK

Wild Profusion: Biodiversity Conservation
in an Indonesian Archipelago
BY CELIA LOWE

Global "Body Shopping": An Indian Labor System
in the Information Technology Industry
BY XIANG BIAO